THE RISE OF SECURITY
and
Why We Always Want More

Mike Croll

Universal-Publishers
Irvine • Boca Raton

THE RISE OF SECURITY and Why We Always Want More

Copyright © 2023 Mike Croll. All rights reserved. No part of this publication may be reproduced, distributed, or transmitted in any form or by any means, including photocopying, recording, or other electronic or mechanical methods, without the prior written permission of the publisher, except in the case of brief quotations embodied in critical reviews and certain other noncommercial uses permitted by copyright law.

Universal Publishers, Inc.
Irvine • Boca Raton
USA • 2023
www.Universal-Publishers.com

ISBN: 978-1-62734-432-6 (pbk.)
ISBN: 978-1-62734-433-3 (ebk.)
ISBN: 978-1-62734-423-4 (aud.)

For permission to photocopy or use material electronically from this work, please access www.copyright.com or contact the Copyright Clearance Center, Inc. (CCC) at 978-750-8400. CCC is a not-for-profit organization that provides licenses and registration for a variety of users. For organizations that have been granted a photocopy license by the CCC, a separate system of payments has been arranged.

Typeset by Medlar Publishing Solutions Pvt Ltd, India
Cover design by Ivan Popov

Library of Congress Cataloging-in-Publication Data
available at https://lccn.loc.gov

CONTENTS

Introduction vii

CHAPTER 1
Basic Instincts 1

CHAPTER 2
Stoking Braziers to Stoking Fear 13

CHAPTER 3
Tough Guys and Rough Justice 31

CHAPTER 4
Security for Sale 43

CHAPTER 5
Generating Anxiety 61

CHAPTER 6
The All-Seeing Eye 75

CHAPTER 7
Hijacks and High Security 91

CHAPTER 8
The 9/11 Bonanza 111

CHAPTER 9
7/7 Perspective Lost 125

CHAPTER 10
Guns for Hire 139

CHAPTER 11
Flak Jackets not Dinner Jackets 155

CHAPTER 12
Enter the Lawyers 173

CHAPTER 13
Heaven or HAL 189

CHAPTER 14
Phreaks, Geeks, and Hackers 203

CHAPTER 15
Homo Sapiens to Homo Securitas 217

Select Bibliography *235*

Web Resources *243*

Biography *245*

Acronyms *247*

Acknowledgements *249*

Index *251*

INTRODUCTION

Security is a baggy term, although its essence is easy to define. Rooted in the Latin *se*—without, *cure*—care, it means without care or worry.

There are many types of security: geopolitical, national, military, economic, financial, emotional, and more. But this book is about physical security: how we protect ourselves from other people. It's not about defence, that's what soldiers do; neither is it about law and order, that's what police officers do. But security is closely related to both and its central character is the security guard who you ignore in the museum, or grumble at in the airport.

Tell anyone that you work in "security," and they are likely to tap the side of their nose and imagine that you belong to some shadowy government agency. Or they might think that you wear a peaked cap and stand outside a supermarket in the rain. Either could be true.

It's rarely been glamourous, but we've needed security since the earliest of times and it's now one of the world's fastest-growing businesses. By 2020 it was worth over $250 billion, close to $500 billion if we include the booming cybersecurity business. Forty countries have more security guards than police officers and in the UK they outnumber police by more than two to one.

The traditional paraphernalia of security was truncheons and lanterns, locks and keys, gates and walls. Today it is metal detectors, CCTV, electronic alarms, X-Ray machines, digital access control, and intruder detection systems. All things that have become so common that we barely notice them.

Yet, increasingly, security systems using facial recognition, GPS tracking, and dataveillance, notice us. We are trading our liberty, and our privacy, for our security. But that may not be a bad thing.

The central question that this book aims to answer is: why do we always want more security? There is no simple answer.

Security is closely linked to safety and, although the terms are often used interchangeably, they have different meanings. Safety also has a Latin heritage: it comes from *salvus*, which means being uninjured or physically unharmed, whereas security is about being psychologically untroubled.

Whatever their etymological origin, safety and security are two sides of the same coin. Both are about protection from harm, but there is a key distinction: safety is about protection from things (trips, tornadoes, and tigers), whereas security is about protection from people (mostly men). Safety is straight-forward and often predictable, but security is more complex because people are endlessly cunning. Armed with sufficient determination, they will eventually overcome any security measures.

The importance of both safety and security was put into context by US psychologist Abraham Maslow in 1943 when he described a hierarchy of human needs. The first are physiological: air, water, food, sleep. Without these, we cannot live. Next, comes safety and security. Without these, we cannot survive. After that comes kinship and belonging to a community, which provide contentment, followed by esteem, dignity and self-actualisation which deliver fulfilment.

Maslow's hierarchy: safety and security are fundamental to the human condition.

Our instincts for security were honed hundreds of thousands of years ago as we competed for survival amongst other animals on the African plains. We were never the fastest, nor the strongest, but we had efficient fear mechanisms that helped us fight, or more likely, flee from predators. Using intelligence and teamwork, we became the apex species, and we developed reason to fear each other more than we feared lions.

As social creatures, we grouped together under an alpha male. He took the best food and the finest women but, in return, he provided security for the group. Protecting people was the first role of the leader, a principle that holds true today.

Communities grew and alpha males became kings. Security became a collective responsibility with all men obliged to take turns at keeping watch and chasing criminals. During the industrial revolution, small towns grew into complex cities. The night watch became a paid service, and police forces were established to maintain law and order.

As people became wealthier, they developed a fascination with locks to protect their growing range of possessions. Locksmiths became famous and lock-picking competitions were celebrated public events. In London, newspapers fuelled a moral panic about crime which boosted demand for locks and security guards, and created a market for home insurance.

Large-scale commercial security started in the US during the 19th century, where big business employed guards to control restive workers, often with fatal results. During the World Wars military guards protected the US's defence establishment. After WWI they were disbanded, but after WWII and the transition to the Cold War, they were privatised and by 1950 there were half a million private security guards in the US.

In the 1960s there was a rapid expansion in the construction of commercial property including office blocks, shopping malls, cinemas, airports, and sports stadia. The owners of these properties, rather than the police, were responsible for security within them, so they turned to private security companies for help.

The 1960s also saw the start of the mass consumer age. Everyone wanted radios, televisions, cars, record players, and kitchen appliances. Some could afford them, others couldn't. Throw in the rise of drug culture and addicts seeking to fund their habits, and the breakdown of traditional family, religious, and community structures, and the result was a crime wave. The main beneficiary was the private security industry which provided protective services to businesses and residences.

Until the 1970s you could board a plane without any identification or baggage checks, and join the pilot in the cockpit for a smoke. Then along came hijacking and the introduction of strict security measures. They slowed things down at the airport, but they helped the commercial aviation security industry to take off.

The age of terrorism coincided with the arrival of graphic around-the-clock television coverage which amplified violent acts and generated public fear.

The 9/11 attacks were like a disaster movie plot. There was an appalling loss of life and everyone felt vulnerable. It also proved to be a bonanza for the security industry which broadened its services to include geopolitical analysis, risk management, employee vetting, crisis management and business continuity management.

Embassies became targets for terrorists, and diplomats were posted to active war zones, swapping dinner jackets for flak jackets. From welcoming symbols of national prestige, embassies were transformed into bomb-proof bunkers. They relied for protection on an exclusive part of the security industry: the private military company, whose hired guns never quite shook off their mercenary image.

In the meantime, lawyers had entered the security arena. Traditionally, if someone suffered a loss in a security incident, they might get tea and sympathy, and perhaps a modest settlement on the side. But by the 1990s lawyers, often on a "no win, no fee basis," were helping victims to fight for substantial compensation. They held venue owners to account for incidents on their property and the fear of litigation led to yet more investment in security.

Cyberspace created a new realm for security. The problem started with hackers who developed malicious viruses to damage computers and soon realised that there was money to be made. They morphed from vandals to criminals and were soon stealing more money than stick 'em up thieves. As organisations became dependent on computers, they were forced to invest in cyber security.

New technology, much of it originating in military systems, found applications in private security: CCTV (first used to monitor the launch of German V2 rockets during WWII), ultrasound and infra-red detection, satellite tracking, magnetic strip access control cards, electronic alarms, and drones. Sleepy guards were replaced by the unblinking eyes of cameras and sensors, controlled from high-tech operations centres.

The security sector has grown rapidly in recent decades. It attracted major investment from private equity which spotted the potential for growth and the need for competent security services to protect their interests in big businesses. The expansion of commercial security has been accompanied by consolidation. In 2021 a series of acquisitions made the security company Allied Universal one of the world's largest enterprises with annual revenues of $18 billion and 800,000 staff. It's a good bet for the future too, as the private security business is likely to grow strongly for many years to come.

It seems that the more security we have, the more we want. Collectively, we are like a donkey trying to eat a carrot on the end of a stick that is harnessed to our neck. As we move forward to take a bite, the carrot remains stubbornly out of range. The gap between our cravings and our satisfaction never closes. We can never be free of worry, so we try to soothe ourselves with ever-increasing security measures.

In part this desire for security is a normal human instinct, but there is also an alignment of powerful interests. Politicians, whose first duty is to protect us. Intelligence agencies, which justify their existence by pointing to unseen threats. The media, which sells more newspapers by shocking us. The security industry, which makes money by selling services to salve our fears. The insurance industry, which compensates for losses whilst minimising the chance of a pay-out. These forces converge on our anxiety-prone minds. It seems, as Dwight Eisenhower said of the military industrial complex, "We will bankrupt ourselves in the vain search for absolute security," and the same could be said of our desire for personal security.

But that, too, may not be a bad thing. There are more feckless pursuits on which to spend our money. Security is a fundamental need for individuals, and it is the first function of government. As early as 1911, German soldier and writer Wilhelm Balck noted that "the steadily improving standards of living, tend to increase the instinct for self-preservation." It's natural that the wealthier we become, the more we have to protect and the more security we demand.

This book is panoramic rather than encyclopaedic. It's an overview, not an oracle. I'm a security insider and I've aimed for a narrative that is interesting and engaging, a sideways perspective on a broad subject. It will take you on a path that may be familiar, but expect some surprises along the way.

Any discussion of security inevitably touches on a lot of injury and death. I haven't lingered on tragedy; others can articulate that much better than I can. I'm not insensitive to suffering but, in this book, I've taken a clear-eyed approach to loss. I don't mean to whistle as I walk past the graveyard, and I mean no disrespect by focusing more on the data than on the human element of some awful events.

Where I've used a quote or a distinct concept, I've given credit in the text. I've avoided footnotes (too distracting) and references (facts can be so easily checked on-line these days), but I have provided a select bibliography so you can see that this isn't entirely a work of imagination. I've used various numbers to make my case. I've tried to be accurate, but do treat them all as indicative rather than gospel because statistics relating to security are notoriously sketchy. I'll take the bullet for any errors.

A brief word on sex. Security, until recently, was an almost exclusively male occupation. I've referred to night watch*men* and I've used the pronouns he/him for security guards. I've done this because it was largely true, and to maintain a brisk narrative without resorting to caveats and contortions. If my instinct for brevity appears insufficiently inclusive, I can only apologise.

This book largely focuses on the UK and the US. These are the countries that I know best, they were amongst the earliest to establish private security, and the US in particular, has been the major innovator in this field.

Why did I write this book? I'm interested in how security developed and where it's heading. I believe there should be more debate about the implications of our insatiable appetite for security, the surveillance culture that has quietly been gathering intensity in recent decades, our attempts to eliminate even the most unlikely risks, and the profound impact of new technology. I aim to illuminate for the general reader how security has become such a major factor in our lives, and to encourage security professionals and students, to think more widely about the subject. Also, I find most of what is written about security to be somewhat dry and inaccessible, so I've tried to be kind to the reader and provide perspective, context, colour, and occasional glimpses of levity.

You might be hoping for some sensitive beans to be spilt on the various organisations for which I've worked. Sorry to disappoint you. You'll find nothing here that isn't available through open sources if you dig deep enough.

I hope you enjoy this story of how Homo sapiens became Homo securitas.

1

BASIC INSTINCTS

The journey to our modern concept of security has its roots deep in pre-historic times. We may be top of the food chain now, but we haven't always been. After descending from trees, we started life as Homo sapiens on the East African savannah around 200,000 years ago. The world was then much richer in animal life than it is today—and as a mid-size mammal that survived by foraging, we were pretty puny. Individually, we still are.

Consider the statistics. Usain Bolt, after years of training, and with designer shoes, can, for a few seconds reach 45kph. That's about the same top speed as an elephant. A wild dog can run at 70kph, and a cheetah can top 110kph. So, there are plenty of creatures that can catch us without struggling for breath.

In a stand-up fight, how would we fare against other animals? Could a 60kg man (that's the average weight of an African man today; the average American man, by the way, is 25kg heavier) overcome a 180kg lion, or even a 30kg baboon? Of course not. We lack the speed, the claws, the teeth, and the aggression, to subdue any but the smallest of animals in a bare-knuckle contest.

Without weapons, traps, and teamwork, we are pathetically vulnerable; so, our finely tuned fear mechanisms are fundamental to our survival. Fear is our emotional response to perceived danger. It's our instinctive risk-assessment mechanism, it's central to how we feel about security, and, as we'll, see it can often defeat clear thinking.

Much as we may like to think that we are predominantly intelligent creatures with emotions, we are fundamentally emotional creatures with intelligence.

Our bodies, our minds, our chemistry, our instincts, are basically the same now as they were 200,000 years ago. We may be better groomed, better fed, better mannered, and smell more fragrant, but we are essentially the same creatures.

The Importance of Being Fearful

Long before we developed language, we were expert at identifying emotions in others. Of the six main emotions generally recognisable across all cultures—fear, anger, happiness, surprise, disgust, and sadness—fear is the one that kept us alive. Fear alerts us to danger and keeps us secure. "Fear", observed Samuel Johnson, "is implanted in us as a preservative from evil."

Life on the savannah was all about survival: getting a meal, without becoming a meal. Mostly that meant foraging for nuts, berries, leaves, insects, and fruit, while watching over our shoulder for predators. If we wanted meat, we had to make do with carrion that had been killed by lions, then picked over by hyenas and wild dogs. After we gnawed on the bones, we left the scraps for the vultures. Our natural place in the food chain is between a canine and a carnivorous bird.

We needed at least 3,000 calories a day to survive, all of which had to be found in the wild. There were no cafes, supermarkets, take-away joints, or neat rows of carrots and corn growing in freshly tilled fields. If we didn't find food, we grew weak. If we grew weak, we fell behind. If we fell behind, we died. We grubbed and gleaned, like the animals we were, and we ate everything raw up to about 10,000 BC, when we learned how to cook with fire.

Very few people in the developed world have had to go without food. We may have felt hungry, but we haven't felt hunger. Hunger is when we don't eat enough to sustain our nutritional needs. It's the point at which the body starts to devour itself, first any reserves of fat, then muscle. Getting enough to eat each day was, until modern times, everyone's major preoccupation, our survival depended on it.

As Homo sapiens, we had a life expectancy of perhaps 25 years, a clear indication that only the fittest survived. Beyond that age, we lacked the speed, strength, and stamina to find food. Contrast this with modern athletes who can compete internationally well into their 30s, and you get a sense of the physical demands placed upon our ancestors. They had to be superbly fit, not to win a gold medal, but just to scratch around for dinner. As English philosopher Thomas Hobbes hypothesised, the natural state was one of "continual fear and danger of violent death and the life of man [was] solitary, poor, nasty, brutish, and short".

At sunset, there was no gentle carefree sleep. Night-time was fright time for our ancestors. Many animals—including hyenas, snakes, scorpions, and leopards—are nocturnal hunters, while the lion is crepuscular, hunting in the twilight of dawn and dusk. The agitation felt by animals around the full moon is the response to the increased predator activity under its silver light. But regardless of the phase of the moon, we have always felt vulnerable at night when our bodies are tired, our vision is reduced, and our predators are active.

Survival on the savannah meant being constantly alert and ready to react. Faced with danger we would flee, or, if cornered, we'd fight. We didn't have time to contemplate, we needed to react instantly. Instinct, not intellect, determined our response.

All creatures are programmed to preserve their lives. It is the most crucial of predispositions. The thing that makes us most fearful is, of course, the prospect of death, especially a painful death. Pain is a natural response to our bodies being harmed. We don't have to ask if we are being hurt because pain tells us. It's a feeling, not a thought, it's an automatic self-preservation mechanism.

Fear, too, is a feeling. We have all experienced it with varying intensity: a sudden knock at the door, a sharp swerve of a car, or an unknown figure bursting from the shadows. Fear explodes deep within our brains in a region called the amygdala.

The amygdala is both powerful and stupid. Powerful because it triggers the release of adrenaline and cortisol from the adrenal glands found above our kidneys. It's like having an espresso, a Red Bull, and a steroid shot all at once. You don't have to think about it, it just happens and instantly you are ready to run or wrestle. Stupid because it creates tunnel vision and blocks out rational thought.

You'll be familiar with the feeling. The thump in your chest as your heart accelerates to pump more blood around your body. Your breath quickens to oxygenate your blood. Your muscles twitch as adrenaline readies them for action. Instantly, you are hyper-alert and goosebumps rise on your skin. You sweat to stabilise your temperature. If you're male, your scrotum tightens as your testicles are lifted, like an aircraft's undercarriage, as you prepare for action. And you get that 'butterflies in the stomach' feeling as blood is diverted from the digestive system into your muscles.

Anxiety is closely related to fear but it's not the same. Fear has immediacy. You see a lion, instantly you dump performance chemicals into your bloodstream, and off you scoot. Anxiety doesn't hit you; it gnaws at you and won't let you rest. Imagine crossing an African plain, with no refuge in sight, and knowing that lions may lurk in the long grass. You don't get a sudden shot of adrenaline, you get a steady drip, like a leaking tap. Your heart canters rather than gallops, your breathing is brisk but not rapid, you are unable to concentrate on anything, and your thoughts keep returning to what might be out there.

These primeval fear mechanisms continue to shape how we feel about security. The term *security* means a state where we are free from fear and anxiety. So, security is fundamentally about managing these feelings which are generated when we believe that we might be harmed.

For most of us the threat of lions has been replaced by the threat of burglars, muggers, or terrorists. But our physiology has not recalibrated to account for our less risky lives. The amygdala remains a commanding and relentless mechanism designed for the dangers of the savannah rather than modern urban environments. It overrides our rational thoughts compelling us to respond emotionally to safety and security issues even though we can expect to live vastly more comfortably, and three times as long, as our ancestors.

Alpha Males

As Homo sapiens, we generally lived in small groups of one or two dozen, and certainly no more than about 150. This was the maximum size of a community in which everyone could know each other. Beyond this number, groups lost their cohesion and sub-groups formed and skulked away. Key to cohesion was the 'alpha male' who kept order, literally through brute strength.

Vulnerable to predators, Homo sapiens stayed together for security, keeping watch, and, like families of meerkats, warning each other of danger. A group of strong males might have the best chance of short-term survival but in the longer term they would die out: reproduction was essential, so they needed a gender balance within the group. Producing and raising children, nurturing and protecting them, was a collective effort. So, security and reproduction, and therefore survival, depended on social cooperation.

Along with an instinct to detect dangerous predators, we also developed a strong sense of whom, within our species, we had to be wary of. The male could immediately gauge how he would fare in a fight with one of his own. He knew in a flash—by his opponent's size, his physique, the pugnaciousness of his features, and the look in his eye—if he could best him in a fight, or if he needed to run away.

That instinct remains with us. When men meet for the first time, they automatically assess each other's propensity for aggression. The clearest guide is facial breadth. A wide face is an indicator of higher levels of testosterone, larger bones, and greater muscle bulk. Richard Wrangham, in The Goodness Paradox, describes how wider-faced men score higher on the psychopathic trait of "fearless dominance". In ice hockey, for instance, the number of minutes spent in the penalty box is higher for broad-faced men than those with narrower faces. In the same way, whenever a man meets a woman, he knows instantly whether he finds her attractive or not. Back in the day, there was little time for an elaborate courtship. They made an instant left or right swipe. Mating was fast and furtive and a brisk performance was a necessity, not a source of shame.

The dominant male in the group naturally became the leader, the alpha male. His life revolved around the three fs: feeding, fighting and reproduction. He got the pick of the females in a social system that favoured breeding by the strongest and fittest, a selection mechanism that played an important role in maintaining the quality of the gene pool and thus the survival of the species.

The alpha male established a principle that has held good through the ages: there can be no political power without physical power. But his privileged position brought with it responsibilities. It was essentially an early example of an implicit social contract described by 18th-century French philosopher Jean-Jacques Rousseau, in his book *Du Contrat Social*. The alpha male had to ensure order within the group and maintain stability. And he had to protect the group from external threats such as predatory animals and aggressive people. If he failed in either of these, he lost his position, and probably his life. In essence, the alpha male's status depended on his ability to provide security for the group. The same remains true of any leadership position.

Foragers to Farmers

A hundred-thousand years ago, when our species perhaps numbered only about 5,000 souls, some of our ancestors walked out of Africa, club in hand, to explore the world. After about 85,000 years we were thinly spread across every continent and had grown in number to about four million, or about the same as New Zealand's population today. To put this in perspective, the world is now growing at a rate of around 80 million people per year—roughly the population of London being born every month.

As we spread and multiplied, we turned our grunts into language, crafted tools, discovered the wheel, and we made best friends with the dog. By 5000 BC we started to grow food rather than chase it, which meant an end to the nomadic life and settling down.

A single hunter-gatherer needed about 150 hectares of favourable habitat to sustain himself. A cultivated plot of that same size could sustain dozens of people. New, permanent settlements sprang up along rivers: the Indus, the Nile, the Yangtze, the Tigris, and the Euphrates. An abundant supply of water, fertile alluvial plains, native cereals, and warm weather created the 'Goldilocks conditions' for sustainable farming which allowed communities to grow well beyond the few dozen members of nomadic groups.

With farming, security took on a new dimension. As nomadic foragers, we roamed the land and only possessed what we could carry. As farmers, we looked ahead, not just to our next meal, but to the next harvest. This meant owning land, building permanent homes, acquiring a range of possessions: tools, draft animals, and furniture, and it also meant stocking food to feed ourselves between harvests. At this point in our development, our major preoccupation shifted from protection from predators, to protection from other people who coveted what we had.

Farming communities came into conflict with two types of people: hunter-gatherers, who had no concept of private property and sought an easy meal, and neighbouring communities competing for resources. The transition from nomadic wanderers to settled farmers brought with it epic conflicts as we staked out the land and attempted to protect our newly claimed property.

Studies of ancient corpses show that hunter-gatherer societies had a violent death rate of 164 per 100,000 per year, compared with 595 per 100,000 per year in early farming communities (by contrast, the worldwide homicide rate in the 21st century is 5.2 deaths per 100,000 and even during WWII it was only around 200 per 100,000—about a third of that in amongst early farmers). We often think of farming being a gentle, noble endeavour, but there is little doubt that it brought with it a significant shift in levels of violence and insecurity.

The Law, the Lash, and the Lord

As farming communities grew into large and complex civilisations, they developed hierarchies. The alpha males' position was formalised and ritualised and they became known as kings. Like the alpha male before them, the king had an implicit social contract with his people. His privileged position depended on him defending against external enemies and maintaining internal order. But with so many more people to govern and protect, he could not do this alone, he needed an army to do his bidding. To support an army, yet more food was needed. More food meant more land, and more land meant more conflict. It was the original vicious circle.

Early civilisations were essentially militarised agricultural societies. To survive they required strict discipline. An alpha male alone could control small hunter-gatherer groups, using his physical strength, and, quite literally, the law of the jungle. Then, as communities settled, the law of the jungle became social norms. As these expanded into large civilisations, the king needed to exert control beyond those that he could personally see and for this he needed laws. Laws provided the benchmark for internal order, they regulated behaviour, protected rights, and provided a framework to resolve disputes. If people stayed within the law, security would prevail. Security provided stability, stability gave rise to productivity, and productivity delivered prosperity.

One of the best-known early sets of laws is from Hammurabi in Mesopotamia, which was written on tablets of stone in around 1800 BC. The ancient Egyptians, Chinese, Greeks, and Romans all developed sophisticated legal codes, based on their notions of appropriate behaviour, and they have become a defining feature of every nation-state. But having laws is one thing, getting people to abide by them is another.

John Locke, the 17th-century English philosopher, believed that everyone had a natural right to protect their lives, limbs, liberty, and property, accompanied by a natural right to punish those who infringed these rights. By *natural rights*, he meant those held by all people prior to the formation of the state. Allowing people to exercise these rights themselves would lead to anarchy and endless revenge, so people surrendered to the state, their right to take justice into their own hands.

Punishments for law-breaking in early civilisations were swift and savage, much, presumably, as they had been in hunter-gather groups. This had two objectives: the first was to penalise wrong-doers, and the second, often more importantly, was to deter wrongdoing in the first place. The deterrence effect was amplified by punishments being a public spectacle.

Serious offences invariably resulted in the death penalty, but it wasn't sufficient just to kill someone, it was felt necessary to make sure that it hurt all the time they were dying. Lawbreakers could expect a slow and painful death: roasted over fire, flayed, or boiled alive, impaled, stoned, or thrown in a pit of snakes. Hanging and beheading were considered mercifully quick, and the guillotine was introduced in 18th century France as a humane means of dispatching the guilty (last used publicly as late as 1939 and not abolished until 1981).

Minor offences attracted corporal punishment. The ancient Chinese used a system known as The Five Punishments, each escalating in severity. First was branding or indelibly tattooing the face; second, cutting off the nose; third, amputation of one or both feet; fourth, amputation of the sexual organs; and fifth, death by quartering or boiling alive. But the most common universal punishment was the lash. It was easy to administer, the number applied reflected the severity of the crime, it was intensely painful although rarely permanently debilitating, and the ferocious crack of leather on bare flesh added to the spectacle. It is easy to forget that until the 1870s the Royal Navy flogged disobedient sailors, and that corporal punishment was only finally banned in British schools in 2003.

Violent punishment was (and in some countries still is) used as the main tool for maintaining social control and security. A sign that a town took security seriously was often the sight of heads on spikes, or crucified or hanged bodies decomposing on the city walls. In Britain, criminals were placed in a body-shaped iron cage known as a gibbet and left to rot. Their flesh would be picked at by birds and eaten by maggots. The remaining bones would be left for years for all to see.

The practice was encouraged by the 1752 Murder Act aimed at "better preventing the horrid crime of murder," by stipulating that, "in no case whatsoever shall the body of any murderer be suffered to be buried." The gibbet was not abolished in Britain until 1828.

As well as laws and the lash, a third mechanism was devised to keep people on the straight and narrow. It was called religion. Hunter-gatherers were animists, believing that animals, plants, and forces of nature had souls. The Egyptians built on these beliefs with animal cults, multiple deities, elaborate funerary practices, and an unshakable conviction that their deeds in this life would determine their fate in the next. And the Pharaohs claimed that they were divinely appointed, which was a master stroke that meant that no one could challenge them.

Most societies developed some form of religion that connected your earthly deeds with your destination after death: the righteous were given a big set of white wings to take them to heaven, but sinners would end up stoking the fires of hell. It was a brilliant concept. Even if no one saw you commit a crime, you couldn't escape God's all-seeing eye (explained more fully in chapter 6). Whatever people thought of their earthly king, they were indoctrinated with a genuine fear of a heavenly God, and this was a powerful means of ensuring order. In the Old Testament, God orders Abraham to kill his son, a clear sign that obedience to God was more important than even parental love. In the US today, studies show that non-churchgoers are twice as likely as churchgoers to be responsible for a crime, clear proof of religion's ability to exert social control and improve security.

Dark Ages

Ancient civilisations eventually lost their cohesion, broke down, and crumbled into smaller states. The impact on security was profound. For example, the Romans occupied Britain for almost 400 years having pacified local tribes and imposed law and order. When they left, Britain was plunged into centuries of anarchy known as the Dark Ages, the rate of violent death increased fourfold, and the economy tanked.

During the Dark Ages, there was no central authority to impose order. Communities had to rely upon themselves for protection using a system of collective security in which everyone had a role to play. Collective security had three key elements: hue and cry, tithing, and the posse.

The hue and cry was a mechanism whereby anyone witnessing a crime would shout to alert others who would pursue and arrest the offender. The practice wasn't unique to the post-Roman Britain; it's deeply rooted in our animal instincts and exists within all cultures. If a hunter-gatherer saw something threatening, such as a snake, a lion, or an aggressor from another group, he would alert others and get ready for flight or fight.

Tithing was a part of a system of compulsory shared responsibility brought to England by the Vikings. A tithe was a group of ten men who were obliged to arrest anyone within their tithe suspected of a crime, or they would all face punishment. This moderated an individual's behaviour by making the group accountable for the actions of each of its members. It was a bit like a football team where if one player commits a foul his teammates are obliged to squeal to the referee, or they would all be yellow carded. It was essentially a self-policing mechanism that built cohesive communities based on cultural norms and mutual trust.

A posse was a contraction of *posse comitatus*, a Latin term for a force of able-bodied men raised to deal with an emergency. It was less immediate than a hue and cry but more organised. A posse hunted for suspects, put down riots, and defended property. It was mobilised by someone in authority, normally someone known as a *Reeve*, who was responsible for an administrative unit or Shire: a *Shire Reeve*, which is where the term Sheriff comes from.

Keeping the Peace

In Britain, after the Norman invasion, the system of collective security disintegrated and much of the countryside became lawless. People gathered in towns for protection, and from the 12th century, an increasing number of them were walled, reflecting the circular relationship between economics and security. To build a wall you needed money, to accumulate money you needed stability, for stability you needed security, for security you needed walls.

Walled towns provided a stable environment, and were the enterprise hubs of the time, but rural insecurity remained a problem. This was the period that gave rise to the legend of Robin Hood. Whatever the veracity of that tale, it was certainly true that Sheriffs struggled to control armed bandits living in the woods.

In the 13th century, King Edward I was determined to "keep the King's peace." This was an imprecise notion meaning tranquility, an ordered state of affairs, an absence of crime, where everyone was secure. This concept has endured through the centuries and even today British Police officers take an oath that they will, "well and truly serve the King in the office of constable ... and ... to the best of my power, cause the peace to be kept and preserved".

In 1285 King Edward issued the Statute of Winchester, which was needed, it stated in its preamble, "Because from day to day, robberies, homicides and arsons are more often committed than they used to be." The problem was that people instinctively supported the indigenous Robin Hoods against the Norman Sheriffs. People would not snitch on their own. The Statute, therefore, aimed to "reduce the power of felons" by enrolling communities. Everyone was given the right to make a citizen's arrest and obliged to make "vigorous pursuit" following a crime. If they failed to do so, they would suffer a "fearful penalty". It didn't specify what the penalty would be, but it would usually involve a combination of extreme violence and debilitating fines.

To make certain that everyone got the message, the Statute was read aloud in "courts, markets, fairs and all other places where people assemble... so that no one can excuse himself on the grounds of ignorance." This was the King losing patience. He was warning everyone that they would be held responsible for criminality in their area.

Under the Statute, tithing was re-energised, and people were obliged to take responsibility for guarding their towns using a system known as "watch and ward." A watch was a watchman, and a ward was an administrative district. Town gates had to be locked between sunset and sunrise, and roads between market towns had to be widened, "so that there may be no ditch, underwood, or bushes where one could hide with evil intent within two hundred feet of the road".

"Every man between fifteen years and sixty [was] assessed and sworn to arms according to the amount of his lands and of his chattels". The statute listed six levels of wealth and the weaponry to be maintained by each. The richest had to have a horse, a chainmail tunic, an iron helmet, a sword and a knife. The poorest had to have a bow and arrow. Everyone was obliged to present their arms twice a year and train with them regularly. This created a militarised society with everyone having a role to play in maintaining security.

Three things point towards the effectiveness of King Edward's system of collective security. The first is archaeological: examination of corpses indicate that the rate of violent death in England halved between 1200 and 1600 which reflected the pacification of the country.

The second is that the system was replicated in America by early settlers in the 17th century and formed the foundation of US law enforcement. The Sheriff and the posse are staples of Western movies which reflected the realities of frontier America where officers of the law would enlist people to help them to impose security or to track down suspects. Their function was the same as those raised in England centuries before.

By the time the Mayflower set sail, firearms had replaced swords as personal weapons, so men were obliged to carry guns and organise themselves into what became known as militias. So the US Constitution with its Second Amendment right to bear arms can trace its lineage back to the lawless forests of 13th-century England.

The third is that the ward and watch, together with the justice administration system, remained virtually unchanged in England until the 19th century when, as we'll see in the next chapter, the industrial revolution, and the complexity of metropolitan life, demanded a new approach to security.

2

STOKING BRAZIERS TO STOKING FEAR

It is sometimes said that prostitution is the oldest profession, but it's likely that watchmen were there first. The earliest written reference is in the Old Testament's 7th century BC Book of Isaiah where watchmen were appointed to guard Jerusalem. And the seriousness of their task is spelt out in the Book of Ezekiel where "If, a watchman sees the enemy coming and does not sound the alarm... I will hold the watchman responsible..."

The watchman is at the heart of the story of security. We can imagine him, a solitary figure in a slumbering town, hunched over a brazier warming his hands, as he kept an eye out for trouble. His direct descendant is the modern security officer sitting in a darkened control room, peering at CCTV images on flickering screens and checking control panels for alarms.

In the first chapter we saw how citizens in England were compelled to mount a night watch, but over a thousand years earlier Rome had a formal internal security structure. Rome was a complex city with a cosmopolitan population of around one million. A city of that size couldn't just rely on laws, religion, and the lash to keep order, and the army was too blunt an instrument for internal use. To maintain security there were three organisations. The Praetorian Guard: elite bodyguards that protected Emperors and high officials. The *Cohortes Urbanae*: which was essentially a police force. And the *Vigiles Urbani*: city watchmen, whose role was similar to that of modern security officers.

Unlike the *Urbanae* they did not enforce the law, rather they prevented it from being broken, and they raised the alarm should fire break out.

Pretorian Guards received one and a half times as much pay as ordinary legionnaires and enjoyed high status. The *Cohortes Urbani* were paid less than a Pretorian but more than regular legionnaires. The *Vigiles Urbanae* were paid less than legionnaires, their work was mostly at night, and their job was so unattractive that it was the preserve of immigrants. Even then, to encourage recruitment and retention, Roman citizenship was granted after 6 years of *Vigiles* service. Some things have changed little. Two thousand years later security officers rarely get more than minimum wage, they often work nights, their status is low, and they are often immigrants who need years of residence to qualify for UK nationality.

Prior to formation of the *Vigiles* in 6 AD there had been a freelance fire brigade formed by General Marcus Licinius Crassus. His brigade would rush to burning buildings, buckets in hand, ready to douse the fire from the nearest water source—once they had negotiated a price from the property owner. If no price was agreed, their place would burn to the ground.

The *Vigiles* were an improvement on Crassus's freelance extortionists, but Rome continued to experience regular fires including one during Nero's rule in AD63, that destroyed much of the city. Across the world, throughout history, fire was a serious and catastrophic risk, with Constantinople, Bremen, Hildesheim, London, Hangzou, Amsterdam, Bern, and Munich, all destroyed by domestic fires before the Middle Ages. Fire control was therefore critical and a watchman's duty included signalling the curfew. In modern times it's an order for people to stay indoors, as we learned during the Covid pandemic, but the term originates from the French: *couvre feu*, literally: cover fire. The curfew was when everyone had to blow out their candles and put out their fires.

Back in England keeping watch had been a compulsory duty since Edward I's time, with every man taking his turn on a roster. As the country become more prosperous and secure, wealthier people, who preferred a comfortable night in their own beds to keeping a look out for fires and thieves, paid for a substitute to do the watch for them. By the 17th century paid watchmen had, in many areas, replaced the volunteer watch.

Watchmen were issued a lantern to guide their way, a rattle to sound the alarm, and a stick to beat intruders. To supplement their modest pay, they helped people home through the dark streets after a night in a tavern, in exchange for a coin. We now take streetlights for granted, but it was 1807 before London got its first gas lights and it was not until 1900 that electric streetlights were common. Other watchman functions included providing a wake-up service for early risers, calling out the time of night, and the unenviable job of removing what was known as "nightsoil" from cess pits and carrying it to dumps outside the city.

Being a watchman was never a glamorous job: outside in all weathers, often solitary, awake whilst others slept, potentially dangerous, their pay was meagre, and their status was humble. In Much Ado About Nothing, Shakespeare created a night watchman character called Dogberry. Self-important, bumbling, often sleeping on duty, Dogberry was a figure of comic incompetence. Like all comedy, it contained more than a grain of truth. In 1698, satirical writer Ned Ward described them as "old frowzy, croaking sots, too infirm and lame to walk without their staves."

It was never a glamorous job.
(Courtesy of Alamy.com)

The watchman would have been someone without land, without a trade or a family business, without money or prospects, and someone who was usually quicker with his fists than with his wits. But blow aside the pejorative, and appealing characteristics emerge. A watchman was a community servant and people relied on him to keep them secure. He needed perseverance and resilience, and he also needed discretion, as he knew about everyone's nocturnal habits. On occasions he needed to be brave, and all the time he needed to be dependable.

Old Soldiers and New Security

Old soldiers often became watchmen. It is what they were trained to do: keeping a look out all night, being loyal, and getting in harm's way if necessary. Few old soldiers had a trade. If they survived their service unwounded, years of hard living would have taken a physical, and perhaps a psychological toll. Military pensions only started in the late 19th Century and even so they were paltry. Their options were often limited to one of the two sides of security: being part of the solution or being part of the problem. Old soldiers made good watchmen. And good criminals.

Concerns were raised about how old soldiers might make a living after the English Civil War, the Wars of the Spanish Succession, and the Napoleonic Wars. They knew how to fight, but not much else. If they couldn't find work, they were well equipped for violent criminality. Gainful employment for old soldiers has long been a preoccupation in post-war environments. In more recent times, it has been a major part of the United Nations' work. Peacekeeping missions in Cambodia, Mozambique, the Balkans, Sierra Leone, Sri Lanka, and Burundi, all had programmes for disarming, demobilising, and reintegrating former combatants. In the UK in 2016 more than 2,500 former soldiers entered the prison system, many of them veterans of wars in Iraq and Afghanistan. So, it's a longstanding and universal issue.

Conscious of the narrow choices facing old soldiers, in 1859 Captain Edward Walter, a retired cavalry officer, and a veteran of the Crimean War, founded one of Britain's earliest security companies, The Corps of Commissionaires. Under the motto, "loyalty, integrity and service," it focused initially on protecting businesses in the City of London. Walter started with eight old soldiers each of whom had lost a limb, including William Turner of the Coldstream Guards who had fought at Waterloo. Immediately identifiable in their smart uniforms, peaked caps, white gloves and rows of medal ribbons, the

Commissionaires' parade-ground presence provided a reassuringly upright welcome for customers, whilst casting an unloving eye upon troublemakers.

The "Original Eight" Commissionaires.
(Courtesy of Corps Security)

The Corps of Commissionaires' social enterprise formula, part security service and part resettlement scheme, was quickly successful and captured the public's imagination, rather like an early version of Help for Heroes. By 1880 it employed over 1,000 old soldiers and branches were opened in Australia and Canada.

The company rebranded as Corps Security in 2008 and remains a significant player in the UK's security market with a turnover of £75 million in 2019. The brass button and parade ground approach has softened in recent years, and it now also recruits civilians and women. Instead of old soldiers standing to attention in draughty doorways, in line with other security companies (as we'll see in later chapters) it went high tech and now delivers many of its security services with remotely monitored CCTV, fire, and intruder detection systems, backed up by mobile response units.

The Corps of Commissionaires stood until the 1930s, as an isolated example of a British security company. While some towns, businesses and wealthy homeowners employed night watchmen on a casual basis, there was little market for organised security services in the UK until the 1960s.

More People, More Crime

For most people in Britain, life had hardly changed from the Dark Ages through to the Georgian period, but industrialisation in the 19th century brought huge economic and social progress The population almost quadrupled from 10 million in 1800, to 38 million in 1900. There were mass movements of people as they were pushed off the land by new, labour-saving farming practices, and pulled into towns by the prospect of factory work.

The nature of society changed as the population grew and became more urban. The traditional mechanisms of social control were loosened. Religion's hold over people's lives slackened and fewer people went to church. In rural communities, everyone knew each other and kept an eye on strangers. In the new towns, almost everyone was a stranger. There was less sense of neighbourliness, and the ancient collective security mechanisms started to dissolve.

As the economy grew, many people became wealthier and started to acquire fine clothes, china, cutlery, lace, watches, paintings, ornaments, and trinkets. But there remained no shortage of hardship and the poor were sometimes driven to steal to survive. Victorian journalist Henry Mayhew noted a correlation, "we increase in poverty and crime as we increase in wealth."

Adding to the dynamic was the Victorians' gradual loss of appetite for the traditional means of deterring crime: brutal public punishment. In the previous chapter, we saw how this had been a fundamental, if unattractive element, of all societies. The first person to challenge the concept was the Italian Cesare Beccaria. His 1764 treatise, *On Crimes and Punishments*, put forward a radical proposal that punishments should be proportionate to crimes, and that law enforcement, rather than the fear of violence, was a better means of social control. He went even further, recommending the abolition of capital punishment, saying the state should not, "In order to dissuade citizens from assassination, commit public assassination".

In Britain hangings had long attracted thousands of unruly spectators. But Beccaria's ideas began to gain traction amongst the educated. In 1849 Charles Dickens wrote to The Times complaining about public executions, "the atrocious bearing, looks and language of the assembled spectators.... screeching and laughing.... with every variety of offensive and foul behaviour.... inexpressibly odious in their brutal mirth or callousness". The last public hanging was in 1868, outside Newgate Prison in London. They continued out of sight, for nearly a century until 1964, when the last person was hanged, but it wasn't until 2004 when the death penalty (for treason) was finally abolished.

Until the Victorian era, the prison population in England was remarkably small, perhaps only a few hundred across the whole country (compared with 87,000 in 2022). Keeping people locked up was expensive and it was considered too light a punishment to act as a deterrent. Most towns had small jails where suspects were held before trial. But it wasn't until 1816, at Millbank, London, that the first bespoke prison, as a punishment and rehabilitation centre, was built. In the following decades dozens more followed, and many from this era remain in use today, including Parkhurst, Pentonville, and Wormwood Scrubs. They borrowed from the French style *architecture terrible*, a statement of brutal austerity, designed to be imposing, and to appease those who felt that prison was too light a punishment. But a key issue remained: how to get the guilty into prison in the first place?

Time for the Police

The problem with Beccaria's theory, that law enforcement was better than punishment, was that, unless someone was caught in the act, they were unlikely to be caught at all. And, if they were caught, proving their guilt before a court of law was time consuming, costly, and often unsuccessful, not least because juries were unwilling to convict people of minor crimes that might result in capital punishment.

Merchants, shop keepers, and wealthy households took care of their own security and hired *thief takers*, essentially bounty hunters, to track down suspects if anything was stolen. Thief takers could be effective, but their use of violence and intimidation made them unpopular. Imagine if law enforcement today was conducted not by police, but by thick necked night club bouncers. The results were often swift, sometimes effective, rarely fair, and never attractive.

Pretty or not, thief takers were a step along the path to modern policing. Henry Fielding, a compassionate magistrate based at Bow Street, recognised that whatever their personal shortcomings, their function was essential for criminal justice to function. In 1749 he funded a respectable brand of thief taker to arrest offenders. They were known as the Bow Street Runners, but, despite their success, there was reluctance to establish fully-fledged police force.

Across the Channel, the French countryside had been patrolled by gendarmes since the 17th century. This was a mounted, quasi-military force, whose name derived from *gens d'armes* meaning men of arms. And Paris had its own military flavoured police which, by the late 18th century, was 3,000 strong.

Because of their centuries-long conflict, Britain was unreceptive to anything French apart from red wine and a white flag. There was also a reluctance to use a military force to do a job traditionally done by community volunteers, and people were suspicious of national institutions.

But the need for law and order in a rapidly growing London was pressing and the Metropolitan Police was eventually established in 1829, absorbing the Bow Street Runners. By 1900, the Met, as it became known, was 16,000 strong (growing to 33,000 by 2020). Determinedly against central control, as the police were established across the country, they were divided into 45 local forces, rather than a more efficient national structure.

The founding father of the British police, Sir Robert Peel, built on the time-honoured principle of collective security by recruiting officers from the communities that they served. His philosophy was that "the police are the public, and the public are the police," and his consensual policing approach continues to this day.

Police officers were paid more than a labourer, but less than a skilled worker. To keep them focussed and impartial they weren't allowed any other source of income, and, until 1887, they were forbidden from voting or showing any political allegiance. Keen to demonstrate a different approach to the army, who wore red at the time and had a reputation for heavy handed suppression, police uniforms were dark blue, and they were armed with a truncheon and a whistle, rather than a rifle and a bayonet.

Locks: Smiths and Picks

As well as prisons and police, the Victorian era saw the industrial production of the oldest of security devices—the lock. Locks had been around for millennia, but in the 19th century reasonably priced, high-quality locks became available to the masses who, in the growing economy, finally had things to secure. Locks were machines in miniature. Cool against the skin, smooth to touch, the key turning evenly within engineered metal was a novel sensation. Locks guarded people's most treasured possessions, they were intimate and exciting, a symbol of progress and achievement, they offered not only security, but also status.

The earliest locks were made by the Egyptians in 2000BC. The first were simple crossbars that slotted across closed doors into brackets on the door frame, literally barring entry. But these had a major drawback—they could only be locked from within, so you still needed at least one guard on the inside. To lock a door from the outside required a leap in technology. It was called a key.

The first key-operated locks were made from timber, using what became known as pin-and-tumbler mechanism. This was a bracket on a door into which slotted a horizontal bolt. A key was inserted into a mechanism that controlled the bolt, lifting a series of pins of different lengths corresponding to notches in the key, allowing the bolt to slide out of the bracket. When the key was withdrawn, the pins tumbled back into place, locking the door. Four thousand years later the classic Yale lock uses essentially the same mechanism. It is one of the most common, and certainly the oldest, piece of security technology.

It is often said that "locks are only for honest people". Even the virtuous might help themselves to unsecured property, but only criminals would intentionally break a lock to steal. Locks establish ownership and control. They are not impregnable, but breaking a lock is a deliberate act with legal consequences.

The Egyptians used locks because they were cheaper and more reliable than guards. Guards had to be fed around 3,000 calories a day. They needed a bed to rest in, and someone to replace them while they slept. In the modern world, staffing a post 365/24/7 requires 5 guards in rotation, with each working 40 hours a week, and taking a few weeks of holiday. In Ancient Egypt, where they probably worked 12 hour shifts without a day off, a 365/24/7 post would have needed only two guards. Even if they were unpaid slaves, they needed to consume 6,000 calories a day between them, or two million calories a year. This equates to about a hectare of grain crops that had to be ploughed, planted, weeded, and harvested and the grain had to be threshed, stored, ground, and baked.

So, while a lock, made from scarce timber, by skilled artisans, was a big investment for Ancient Egyptians, it was still cheaper in the long run than guards. And it didn't fall asleep, or need feeding, or somewhere to rest, it didn't take bribes, or fill its pockets with grain or treasure. You could trust a lock; a lock was the perfect guard.

From Ancient Egypt, locks spread throughout the world and were made of iron, steel, and brass. They grew in sophistication, but their use was largely confined to royalty, the church, and to wealthy families, as they remained expensive and only the rich had possessions that were worth securing. Most people simply secreted any valuables about themselves, or buried them in their hearths.

In the Victorian era, industrial means of producing locks met mass market demand for them. To win publicity for their products, companies staged lock picking contests with their rivals to prove whose products were best. These were like grandmaster chess competitions, and they fascinated a generation.

Lock smiths and lock pickers became household names, and they had a powerful influence on the public's perception of security.

In 1817 thieves used counterfeit keys to open a lock at Portsmouth Dockyard to steal ships' stores. In response, the British government announced a competition for a lock that could only be opened with its own key. Jeremiah Chubb, a local locksmith, won the £100 prize with his *detector lock*. It was supposedly unpickable and had a lever that tripped, alerting the key holder that it had been tampered with, hence the name.

Proud of his achievements Chubb advertised his lock with the words, "Look on my works, ye burglars, and despair." His products were so successful that he became the sole supplier of locks to the Post Office and the newly created Prison Service. Chubb became famous, and, at the pinnacle of his career, he was awarded the contract to provide secure cabinets for the Koh-i-Nor, the world's largest diamond, when it was displayed at the 1851 Great Exhibition in London.

The Great Exhibition was a showcase for industrial, scientific and cultural achievements. As a world leader in secure technology, Britain's locks were prominently on display. Chubb proudly challenged anyone to pick his detector lock which had been impregnable for the 33 years since its invention. The dramatic event became known as "The Great Lock Picking Controversy".

It was a waistcoated American, Alfred C Hobbs, who took on the pride of Britain's locksmithing establishment. After knotting his brow and fiddling for 25 minutes there was a sharp click followed by gasps from the astonished spectators, as the lock opened. Chubb was mortified and the press splashed the sensational story. He went back to the drawing board and salvaged his reputation by modifying his lock with additional anti-tampering features and offering to retrofit other locks that he had supplied.

Alfred Hobbs was no fresh-faced novice. He had spent years selling locks for an American firm called Day and Newell. His sales technique was to examine a competitor's lock in front of its owner. His favourite line was, "There is something the matter with your lock." The reply was inevitably, "What is it?" To which Hobbs, always a showman, would say, as he snapped open the mechanism, "It won't keep the door shut." At which point he would invite them to buy from Day and Newell.

The wily Hobbs went on to test his skill against a challenge lock, which had not been picked since it was patented by Joseph Brahmah in 1790. The lock was on display in the window of the company's shop at 124 Piccadilly, accompanied by the inscription: *The artist who can make an instrument that will pick or open this lock shall receive 200 guineas the moment it is produced*. It was a tougher proposition, but after 51 hours of effort, over several days, Hobbes claimed the prize. He stayed on in London and set up his own successful locksmiths, Hobbs and Co. The company bearing his name continued to flourish even after he returned to the US in 1860.

The Brahmah Lock, patented in 1790, on display in London's Science Museum.

The lock picking competitions had a lasting effect. As well as building brand identity, they also aroused public interest in crime. The Google ngram function, which measures word use over time by scanning historic texts, shows that the word *burglar* jumped 15-fold between 1851 and 1890. Much of this is likely the result of promotions by the commercial lock industry as well, as we'll see, by the press.

Most locks sold across the world today are based on designs by Chubb, Brahmah, Hobbs, and Yale. Their companies are still in business, apart from Hobbs' which had its brand name extinguished when it was bought by Chubb in 1954. Both Chubb and Yale, went on to became massive international brands before being bought themselves in 2000, by the Swedish conglomerate Assa Abloy. Brahmah remains independent and continues to offer, from its offices in London, unpickable locks for high-end clients.

All these years later the public lock picking competition has a modern equivalent. Tech companies will pay bounties to hackers who identify security issues. Apple for example, will pay up to $200,000 to anyone who can extract data protected by their Secure Enclave technology, and Microsoft will pay $250,000 for anyone who can identify critical vulnerabilities in their systems. As we will see in chapter 14, the Victorian lock picker morphed into the geeky hacker.

Defining Responsibilities

Between 1800 and 1900 the number of people living in London increased from 1 million to 6.5 million. As it expanded, The City—bounded by the old Roman walls—became a centre of banking, finance, and insurance. Businesses displaced the residents who headed out to the suburbs to become commuters on the new trams, trains, and tubes. Traditionally people had lived above their business, so there was always someone on site to keep an eye on things. But the depopulation of The City meant that many commercial properties were empty after office hours, relying for security on shutters, bars, safes, a handful of watchmen, and the occasional police patrol.

In February 1865 there was a dramatic robbery in The City. A gang of six thieves spent a weekend cutting through a neighbouring building into Walkers' Jewellers on Cornhill. They forced open the safe, and escaped with £6,000 worth of gems. They were later arrested, charged, and sentenced to 14 years in Millbank prison. It was the UK's largest ever theft at the time, equivalent to more than £800,000 in 2020.

The robbery had remarkable similarities to the Hatton Garden safe deposit robbery of 2015. Just two kilometres from Cornhill, it was also by a six-man gang. The press named them the *Diamond Wheezers* as they were all in their sixties and seventies. Despite their age, they spent the weekend drilling through the vault walls and broke into hundreds of safe deposit boxes. It too was the largest robbery of its time, netting £14 million. Like the Cornhill gang, the Wheezers were caught and convicted.

The Cornhill robbery gripped the nation. The Times reported that, 'For years no occasion of this kind has caused more excitement'. It was another blow to the lock-making industry as the safe, a Milner & Son 'Holdfast' model, was warranted secure against fire and theft. However, the safe's lock had not been picked. A number of small wedges were hammered in between the door and the frame providing sufficient room for a jemmy to prise off the door. Walkers' sued Milner & Son but the case was dismissed. The judge said that a safe alone was no panacea for theft, and that it had to be combined with "care and watchfulness".

The police were criticised for not preventing the robbery. They responded saying that they "cannot be responsible for what may be occurring out of their sight, within deserted buildings, to which they have no access". They were supported by The Daily Telegraph which believed that the police "must be auxiliary to private vigilance, not a substitute for it." The City Press reinforced the point, "if every Englishman's house is still to remain his castle, it must be fortified and manned by himself."

The Cornhill case defined the fundamental responsibilities and limitations of each party, that hold good today:

- Owners have responsibility for the security of their own possessions.
- Security product manufacturers are not liable for theft.
- Police responsibilities do not extend to securing private property on a private premise.

The onus was placed firmly on owners, not on the police. If owners wanted security, they would have to pay for it. This cleared the ground for the take-off of the private security business. The immediate response from businesses on Cornhill was to create a local watch with a detachment of the recently formed Corps of Commissionaires.

Moral Panic

Charles Dickens wrote about low life and criminality in Victorian London. *Oliver Twist*, published in 1839, shocked readers with what today we would call gritty social realism, describing thieves, pickpockets, prostitutes, and house-breakers. Dickens wrote to illuminate the plight of the poor, but he generated anxiety rather sympathy in his readers, convincing Londoners that a criminal class stalked the gloomy streets, preying on the respectable.

Crime, or at least the arrest rate, appeared to be rising; the number of people tried for burglary in England and Wales had increased from 985 in 1843, to 1,625 in 1849. But the figures were hardly dramatic. Unlike homes in France where windows had shutters and bars to guard against theft, these basic measures were rare in Britain and windows and doors were easy to force open. This surely was an indication that burglary had not been uppermost in people's minds. Yet the mind is a productive instrument for the propagation of imagined fears.

A popular writer of the era, George Cruikshank, wrote a pamphlet in 1851, entitled "Stop Thief: Hints to Housekeepers to prevent Housebreaking". He aimed, he said, not at "creating alarm", but "to give such a feeling of security that even nervous persons may lay down their heads up on their pillows at night without apprehension of damage to their property or violence to their persons". Naturally, it did the opposite.

Dickens' criminal characters, Fagan, the Artful Dodger, and Bill Sykes, were already deeply etched into public consciousness. Chubb and Brahmah's impregnable locks had been breached, as had the Cornhill "Holdfast" safe. Now here was Cruikshank explaining how easy it was for people's homes to be burgled. The conditions were ripe for one of the earliest moral panics.

A moral panic is a widespread fear that something threatens society, provoking a disproportionate response. Recent examples include terrorism, AIDS, drugs, dangerous dogs, paedophiles, contaminated sweets at Halloween, the Y2K bug, and the 2021 fuel crisis. Sociologists Erich Goode and Nachman Ben-Yehuda described the five characteristics of moral panics as:

Concern	– that something will have a negative impact on society.
Hostility	– towards those responsible who then become demonised.
Consensus	– about the danger posed.
Disproportionality	– of the action taken against the threat posed.
Volatility	– of public interest which tends to come and go quickly.

The moral panic in Victorian London was about mugging—robbery with violence—or as it was known in the 19th century, garrotting. The panic appears to have emerged from a debate about reforms of the penal system.

The transportation of criminals to Australia was sharply declining at a time when, to ease overcrowding in prisons, many prisoners were released on early parole. But old attitudes to punishment lingered. Prison had only been used as punishment for two or three decades. Many still saw hanging as the best option, transportation a possible alternative, and prison as too soft. Early parole, many thought, was an outrage that would inevitably lead to more crime.

In November 1856 the British Prime Minister, Lord Palmerston, announced that Britons should "feel safe to travel the world". Taking issue with this, the editor of The Times reflected a common notion that even in London there was, "imminent danger of being throttled, robbed and if not actually murdered, at least kicked and pommelled within an inch of his life". Crime statistics from the era are notoriously patchy, but of the 6,762 burglaries recorded in England and Wales between 1843 and 1849, only 45 involved violence. This hardly represented mortal danger. However, rarely pausing to let the truth get in the way of eye-catching drama, the press brewed up a storm, leading with the story day after day, the product of which was a moral panic.

The consequence of this was more heavy-handed policing, longer sentences for criminals, less parole, and a market for anti-garrotting neck collars. Gunmaker Henry Ball even invented the "anti-garrotter belt pistol." Worn on a belt at the small of the back, it had a flat plate rather than a conventional handle, a firing mechanism operated by a string, and a single shot aimed at hitting a garrotter in his midriff.

The panic harnessed vested interests: the press, who could sell more newspapers; the politicians, who wanted to be seen to be tough on crime; the police, who could leverage the panic to increase their numbers; the security companies, who could sell more locks and hire out more guards; and of course, the public, whose fears had been confirmed. The alignment of these interests has played out repeatedly ever since with emotion inhibiting rational discussion about security.

Selling Fear

Wealth provides material comfort, but also plants a sense of unease that your possessions might be lost or stolen. We have a tendency known as *loss aversion* where we become more preoccupied by loss than by gain. Losing £10 generates more negative feelings than the positive ones generated by winning £10. So, we are instinctively loss averse. This phenomenon probably has an evolutionary basis. Our fear of loss generates adrenaline to help us fight or flee, and it's fundamental to our survival. The prospect of gain generates dopamine, which makes us feel pleasure. For example, if we are eating a chocolate cake, and we hear an explosion, we will immediately leave the cake and seek safety. But if we are running from an explosion and we see a chocolate cake, we don't stop to eat it. Our instinct for survival will always be stronger than our desire to feel good. Adrenaline beats dopamine.

In the 19th century, the new middle classes were anxious to protect their possessions, especially having been spooked by the media into believing that, at any moment, they could lose all that they held dear. This was fertile ground upon which to sow commercial seeds. Never slow to chase an opportunity, the insurance industry spotted a market for burglary products. Whether crime had been rising or not was irrelevant, for the fear of crime had seized the public. It was the perception, not the reality, that counted—the feeling rather than the thought. For insurance companies, the wider the gap between perception and reality, the bigger the gain. The public would pay a premium to reduce a threat perceived to be highly likely, whilst the insurer had to compensate for loss that was in reality highly unlikely.

The garrotting panic resurfaced in 1862 following the actual mugging of an MP on Pall Mall. The press again whipped the public into a frenzy with The Times alone publishing 18 editorials on crime in the following two months. The fear was reinforced in the 1870s by the case of the notorious burglar and murderer, Charles Peace, who was hanged for his criminal deeds. People queued to gawp at his wax effigy in Madame Tussaud's for many years afterwards.

Riding the wave of popular fascination with crime, in 1887 Arthur Conan Doyle created the fictional detective Sherlock Holmes. In the 1890s, E.W. Hornung introduced the public to the fictional gentleman thief A.J. Raffles. Then Agatha Christie and Dorothy Sawyers became celebrated crime writers in the 1920's and the genre has been a staple of public entertainment ever since.

To capitalise on the anxiety, a handful of specialised burglary insurance companies targeted wealthy and middle-class households with more precision than a thief in the night. Their technique was to assert a general feeling of insecurity, knowing that describing fearful emotions in unspecified others would generate the same feeling in their audience. Their message was: *other people are afraid; you should be too*. This nags away at even the most rational. Advertisers keep chipping at emotions to get a sale. It is a classic stoke and soothe technique. Stoke the fear of theft, and then soothe it by providing a loss prevention product.

The General Burglary Insurance Association used the tagline "locks, bolts and bars soon fly asunder." The message was clear, it doesn't matter how much you secure your home, it is still an easy target, so you need insurance. Some companies posted sample policies through letterboxes following a burglary in the area. Others advertised in newspapers with slogans like, "Prudential Insurance Dispels Anxiety", or "the wise businessman and householder will insure", and the more subliminally sinister, "What's happening at home?".

By the 1930s burglary insurance was sold as part of broader home insurance policies which covered a wide range of risks including fire, flooding, damage, and subsidence. Within 40 years almost every home had insurance.

This marked the start of a symbiotic relationship between insurance companies and security companies. The insurance company talked up the risk of burglary, whilst reducing the likelihood of a pay-out by insisting on security measures such as additional locks and window bars, which security companies were happy to provide. As a result, householders were less likely to be burgled, and would be compensated if they were. Once a policy was taken out, it was inevitably renewed year after year. It was a brilliant business model.

3

TOUGH GUYS AND ROUGH JUSTICE

In Britain by 1900 there was an established police force, but although there was a long tradition of nightwatchman, the private security market remained small and largely informal. There were old soldiers and various watchmen who were employed by individual businesses or homeowners, but there was little market for private security companies. By contrast, in the US, by 1900 private security companies offering a range of services were well established, and by 1950 some estimates put the number of private security guards at close to half a million. In this chapter we'll look at the development of private security in the US and the creation of its main law enforcement agencies, through four main phases: the first settlers, the settlement of the West, the industrial period, and the World Wars.

Settler Security

Amongst the earliest settlers, arriving in 1620, were the Pilgrims, a new breed of protestants, pious, tight knit, hard-working, politically innovative and egalitarian. They established what was essentially a church state, outlawing festivities, gambling, dancing, and sport, and they showed a marked intolerance of other religions.

They brought with them from Britain the collective security system that had remained virtually unchanged since the 13th century: watch and ward, hue and cry, the sheriff, the posse, and the requirement for all men to maintain arms. The major difference was that by the 17th century, as we saw in chapter 2, arms meant guns rather than swords.

During the voyage to America the Pilgrims signed the *Mayflower Compact*, which affirmed the principles of collective security and bound them to "combine ourselves together into a civil body politic, for our better ordering and preservation". This permanently flavoured American life, laying the foundation of democracy, and creating a citizen militia with each man providing his own weapon and wearing his working clothes rather than a uniform.

The natives they encountered were hunter gatherers. They cultivated some crops, but their tools were made of stone, flint, or bone. They had not discovered metal, mathematics, or developed written language. The settlers were technologically thousands of years ahead. They had great sailing vessels, steel tools, mirrors, books, alcohol, navigational instruments and, most importantly, they had guns. Native Americans, like all hunter gathers were supremely fit, and well adapted to survive in the wild, but they were always going to come off second best against firearms.

Despite the difference in technological sophistication, there was less distinction in their moral progress. William Fitzhugh Brundage, in his book *Civilising Torture*, describes a conversation in 1637 between Jesuit missionaries and Huron warriors. The Huron had captured a man from the Iroquois tribe with whom they were at war. Over the course of a day and a half the prisoner was slowly roasted over a fire. The missionaries objected to the torture. The Huron asked if Europeans killed prisoners. The missionaries conceded that they did, "but not with this cruelty". "Do you never burn any?" the Huron probed. "Not often", the missionaries replied, adding that "even then, fire is only for enormous crimes, and besides, they are not made to linger so long—often they are first strangled, and generally they are thrown at once into the fire, where they are immediately smothered and consumed".

The settlers had a precarious start in North America but gradually their numbers grew. New settlers were allocated land in front of the original colony. Like a growing onion, the outer layers provided protection for those on the inside. Being on the outside, on the frontier with native territory, they formed a defensive line until the next boat load of settlers arrived. Although there was cooperation and trade between them, most settlers regarded the natives as barbaric godless savages and had little compunction about seizing their land, so the frontier was always a zone of conflict.

In 1620 there were perhaps 5 million natives in North America grouped into 600 tribes. By 1800 the native population was down to 600,000, falling to 250,000 by 1900. Perhaps 30,000 died in conflicts with settlers, however, the main cause of the precipitous decline were diseases such as chicken pox and measles introduced from Europe. By comparison, the settler population rose from 5.3 million in 1800 to 76 million by 1900.

Militias had been key to settler survival and expansion, and they gained legendary status in 1776 by winning the war of independence against the regular British army. The 2nd amendment to the Constitution cemented their role and that of weapons into the US law stating, "a well-regulated militia, being necessary to ensure the security of a Free State, the right of the people to keep and bear arms, shall not be infringed." Guns have been a key aspect of Americana ever since.

The Wild West

The British had forbidden settlers from moving beyond the Appalachian Mountains as it was too challenging to govern and tax them. But after Independence, first a trickle, then a flood of people headed West. Some saw this as fulfilment of their "manifest destiny" to bring civilization from "sea to shining sea," others just wanted land or gold.

The gun was the main tool for settling the West. You needed one to shoot animals for food, to kill predators such as bears and wolves, to fight off natives, to protect your family, to defend land, to be a member of a militia, and to show that you were a man. The new industrial age brought mass produced, affordable and reliable weapons including the Colt revolver and the Winchester repeating rifle, which was celebrated as the "the gun that won the West".

With huge tracts of land being parcelled up, often on featureless terrain, the problem of creating boundaries and preventing cattle from escaping was solved by the invention of barbed wire. It worked equally well on people, and it has been used in vast quantities ever since for military defence, prison security, marking international boundaries, and for protecting property. Barbed wire also has such strong symbolism as an instrument of oppression that Amnesty International's logo is a candle surrounded by it.

The West was a lawless place where might was right and possession was nine tenths of the law. In the absence of banks, post offices or shops, there was a demand for secure transportation for cash, mail, and supplies which was fulfilled by what were known as express companies. Using horse drawn stagecoaches with strong boxes and armed guards "riding shotgun," they supported settlers travelling from the East to the frontier lands out West. In some ways it was a similar service to that provided by the Knights Templar to the Crusades in the 12th and 13th centuries.

In 1850 freight agents Henry Wells, William Fargo, and John Butterfield formed a company in Buffalo, New York, known as American Express. They argued about the westward expansion of the business. In the end American Express confined itself to lands East of the Missouri, while in 1852 Wells and Fargo named a new company after themselves and went West. Both companies flourished and became world-wide brands. American Express eventually concentrated on finance, but Wells Fargo also retained its secure transport business. Its trademark red and yellow armoured cash in transit vehicles remain a common sight across the US.

Another early secure transport company that survives today is Brinks. It was started in 1859 in Chicago by 29-year-old Perry Brink with a single horse-drawn carriage transporting people and luggage between rail stations and hotels. From there he, too, moved into the secure transport business. Brink guaranteed delivery or compensation for any lost or damaged shipments, which was a masterstroke in developing trust and is an enduring part of the cash in transit business model. Nowadays cash in transit is almost exclusively a contract business. Everything is insured by the carrier, so the customer has no liability for any losses. Brinks flourished and in 1904 he traded horses for motor vehicles and then moved into armoured trucks. Rapidly expanding across the country, Brinks went global in the 1960s and by 2010 had more than 130,000 staff in 100 countries.

There were three broad phases of securing the West: vigilante justice, community justice, and finally the establishment of professional law enforcement bodies. Theodore Roosevelt put it this way, "As soon as the communities become settled and begin to grow with any rapidity, the American instinct for law asserts itself, but in the early stages each individual is obliged to be a law to himself and to guard his rights with a strong hand." Those heading West were a different breed from the pious settlers of the East Coast. Many were young men, full of hope, ambition, and testosterone, seeking adventure and fortune. Their down time was not spent in church or studying the bible, but drinking, gambling, and fighting.

Guns, competition for land, the absence of established cultural norms, and a prickly sense of honour, meant that disputes quickly turned violent. Hanging, flogging and banishment were commonly used to punish offenders following a quick show of hands by a hastily convened court. For example, in 1851 in San Francisco a *Committee of Vigilance* believed that Australian immigrants were responsible for much of the city's crime. They deported many and publicly hanged four men accused of murder. This was a fast and hard way of reducing crime, and it had popular support.

Enforcing the Law

When the law finally asserted itself, it came in the form of Sheriffs and Marshalls. The US Marshall Service was the first Federal law enforcement agency founded in 1789, as part of the Department of Justice. The Marshall's main role was, and remains, to execute warrants for arrest, in a similar fashion to the Bow Street Runners in London.

Sheriffs were elected locally by the town's people, so they were highly accountable. Yet it was an ill-defined post and their terms of reference were little more than "keep the peace and run the jail". They had no formal training or prescribed procedures. The saying out West was "there is more law in a Colt six gun than in all the law books". Unlike Marshalls, few received a salary, and they supported themselves by taking a cut of fines, collecting bounties on wanted men, or holding second jobs. It was the ultimate alpha male community volunteer post.

The Sheriff was a tough guy with a gun and strong sense of what was right. By the mid 19th century in Britain, the Sheriff had a largely ceremonial role, turning out at important occasions wearing court dress: a dark blue velvet coat with brass buttons, a frilly collar, breeches, and a cocked hat. In American popular culture, John Wayne epitomised the rugged law enforcement officer, facing down outlaws, gun in his hand, saying, "a man's got to do, what a man's got to do." By contrast, in British popular culture, Sherlock Holmes in a tweed suit, waving a pipe would say, "elementary my dear Watson". Small wonder that Americans sometimes find their British counterparts effete and eccentric.

The Sheriff could appoint Deputies to assist him, and raise a posse to chase suspects. Being part of a posse was a civic duty, like jury service—only more manly. Most people joined in enthusiastically, relishing the opportunity to buckle up a gun belt, ride with the boys, and track down outlaws.

In the US Sheriffs remain a key part of the law enforcement establishment with more than 3,000 across the country mostly in areas with populations too sparce to support a municipal police force. Their role include jail administration, court bailiff, and tax collection, and they are still elected locally like their Wild West predecessors.

The Sheriff—still a tough guy with a gun.

A square jawed man on a horse could enforce rough justice in the open lands of the West but the rapidly expanding cities (for example, New York's population rose from 80,000 in 1800 to 3.5 million in 1900) needed a different approach.

Municipal police forces were established and organised along similar lines to those in the UK, starting in Boston in 1838, and by the end of the 19th century they were widespread. In common with the UK, they remained locally administered and today the US has over 18,000 separate police forces. But unlike the UK where 95% of police officers are armed only with a truncheon, US police officers must assert their authority over armed citizens, so they all carry guns.

In 1823 the notoriously uncompromising Texas Rangers were formed to protect settlers from native Americans. State Police, sometimes known as State Troopers, who were introduced across the US in the early 20th century, modelled themselves on the Rangers and they enforce the law in the gaps between the municipal police and the rural sheriffs, which largely means patrolling highways. With their smart shirts, Smokey bear hats, Sam Browne belts, high boots and jodhpurs, they are a curious reminder of the old days on the frontier.

Given its pioneering settler, war of independence winning, wild frontier heritage, it's inevitable that the US takes a robust attitude to law enforcement. A common symbol of justice, featuring at the House of Representatives, the Senate, the Lincoln Memorial, in the Oval Office, and from 1916 until 1945 on the ten-cent coin, is the ancient Roman *fasces*, a bundle of sticks with an axe in the centre. The Romans used the sticks for corporal punishment (from the Latin *corps* meaning body), and the axe for capital punishment (from the Latin *caput* meaning head), in other words, beheading.

The *fasces* gave its name to fascism which stands for authoritarianism, discipline, and regimentation. It was not a political philosophy embraced by the US, although the US has always taken an uncompromising approach to justice. Judicial corporal punishment in the form of the lash was last used in 1952. Capital punishment was by hanging or electric chair (and since 1977 by lethal injection) rather beheading with an axe. An outlier amongst western democratic states, the US retains the death penalty to this day. The US also has the highest rate of incarceration in the world, ten times higher than in the UK.

The Industrial Age

By the mid 19th century, the industrial age had taken off in the USA. It was fuelled by abundant raw materials, new technology, entrepreneurial spirit, and rising demand for material possessions from a growing population swelled by new waves of immigrants. Between 1870 and 1900, 12 million people answered the call, "Give me your tired, your poor, your huddled masses yearning to be free".

Freedom usually meant drudgery in factories and mines, or on farms or construction projects, living in spartan company accommodation and paid in company currency, in effect they were modern versions of indentured servants. This was the era of American *robber barons* including Vanderbilt, Carnegie, and Rockefeller. Men who became fabulously wealthy on the spoils of the land, and the toils of the workers.

Bringing together huge workforces of immigrants from different cultures was a herculean task that had rarely been attempted before. Settling America had been an epic challenge and the new industrial endeavour was pioneering on a similar scale. The workers needed to be trained, organised, disciplined, fed, accommodated, prevented from thieving, and discouraged from striking. Balancing a commercial proposition with the reasonable treatment of workers was a long struggle.

Disputes over pay and conditions routinely turned violent. Armed vigilantes or militias were used to suppress striking workers resulting in hundreds of deaths between 1850 and 1940, including 40 railroad workers in Pittsburgh in 1897, 20 miners in Pennsylvania in 1910, and 100 farm workers in Arkansas in 1919.

Labour Unions were formed to promote workers' rights using the slogan "an injury to one is an injury to all". They organised a series of strikes that created demand for the first private security companies. One of the earliest was formed in the 1850s by Scottish emigree Alan Pinkerton whose detective agency used the motto "we never sleep".

In 1861 he provided security for Abraham Lincoln en-route to his inauguration diverting the President-elect from a supposed assassination attempt. Debate continues about the veracity of the threat. Either way, it was good for business and Pinkerton's became the most prominent security company in the US. His logo was an all-seeing eye, representing the company's ability to insert detectives within the workforce to identify troublemakers. The eye logo inspired the term *private eye* as a nickname for detectives. Other services he offered included protecting company property, providing bodyguards for bosses, tracking down criminals and breaking strikes. By the 1890s Pinkerton had 2,000 staff and a further 30,000 reserves to call upon.

In 1892 a strike at Dale Carnegie's Homestead Steelworks in Pennsylvania was a pivotal moment in the struggle between labour and capital. Hundreds of Pinkerton guards were called in to break the strike by members of the Amalgamated Association of Steel Workers (known as the AA). The confrontation soon became a pitched battle with both sides using guns, cannons, and firebombs. Ten people were killed (7 strikers and 3 Pinkertons), dozens were wounded, and more than 300 Pinkerton guards were captured and beaten after surrendering to striking workers. Order was eventually restored by

4,000 militiamen. A few days later an anarchist, with no connection to the AA, shot and wounded the steel works manager. This undermined support for the union and the strike eventually collapsed.

The episode reflected badly on the unions, Pinkerton, and on Carnegie. The AA's credibility disintegrated, and membership seeped away. Reflecting public outrage at Pinkerton's actions, the US Government passed an Act in 1893 stating that "An individual employed by the Pinkerton detective agency, or similar organisation may not be employed by the government of the United States". Although Pinkerton's reputation suffered, they survived commercially and continue to trade under the same name (although they were bought by the Swedish company Securitas in 1999). Carnegie went on to restore his reputation by becoming a major philanthropist.

Other big names in private security were created around this time. In 1909, William J. Burns founded a detective agency that within a year took over from Pinkerton as the American Bankers' Association's security provider to 12,000 member banks and continued to expand rapidly, despite occasional scandals. In 1916, Burns was hired by J. P. Morgan, to investigate the theft of trade secrets by a rival company, Seymour and Seymour. He installed a primitive listening device in the firm's offices and wire tapped their telephones, but was caught and fined $100 for illegal entry.

In 1927 Burns was jailed for a fortnight after he arranged for his agents to follow jurors in a corruption trial. His conviction was later reversed, but his reputation for unconventional methods stuck. Like Pinkerton, Burns provided toughs and spies to big business and the two companies dominated the private security business in the US. Burns also remains a major player although they too were acquired by Securitas in 2000.

Rather than contracting private security companies, some big businesses preferred to develop internal security departments. There were advantages to both approaches. Outsourcing was a flexible way of bringing in expertise and capacity whilst an in-house team was more aligned with company values, more loyal, and more discrete. Henry Ford, the car manufacturer, was an early adopter of in-house security.

Ford launched his Model T in 1908. His revolutionary mass production enabled economies of scale and it sold for half the price of competitor models. Mass production demanded mass labour and Ford employed around 14,000 workers mostly on repetitive tasks on assembly lines. In the restive labour environment of the times, turn-over was high and loyalty was low. To maintain a stable, well-disciplined work force, Ford created a Sociological Department with inspectors that checked on the cleanliness of workers' homes, the school attendance of their children, their savings records, and their alcohol consumption. Ford was a pioneer in the development of corporate culture and believed that a healthy workforce was essential for an efficient business.

As well as a Sociological Department to encourage virtue Ford also created a Service Department to prevent vice. "Service" in this case, was a euphemism for security. The Department was run by Harry Bennett, a diminutive streetfighter and former sailor, with a feet-on-desk style, who recruited brawny young men to his team which grew to be 3,000 strong. Legend has it that Ford asked Bennett only one question when they first met: "can you shoot?" The two become close. "I got things done," said Bennett, "That's why Mr Ford liked me". Bennett's team monitored workers, protected executives, guarded factories, conducted investigations, and acted fast and hard against Ford's nemesis—the unions.

In 1932, during the great depression, car workers demonstrated at the Ford plant in Michigan carrying banners saying, "We Want Bread Not Crumbs". Things turned violent and the police and Service Department members shot dead five marchers and wounded many more. Bennett was seen firing a pistol into the crowd. A grand jury investigation concluded that the police "might have been more discreet and better considered before they applied force".

Bennett kept his job but was in the news again in 1937 when his team beat up union officials campaigning for more pay. The event was captured by a press photographer and it became a national scandal. Ford's image may have been dented but it was just a reputational fender bender. Bennett eventually retired in 1945 to a remote ranch in California. Legend has it that his home, in addition to gun turrets and other security features, was fitted with a stair-case that had uneven steps to trip unsuspecting intruders.

World Wars

During WWI the U.S. government guarded key installations, government facilities and munitions plants. The task might have fallen to the militias, but they were enrolled into the regular army to fight in Europe, so the United States Guard (USG) was formed. This was a uniformed military force comprising of 28,000 mostly former soldiers and police officers. They too were involved in supressing strikes, and they were not renowned for their soft-hearted approach. After the war they were quietly disbanded.

WWII demanded much more security for facilities across the US. There was intense fighting across both the Pacific and the Atlantic fronts and a risk of enemy landing parties. Military bases needed guarding and there was concern about sabotage at weapon production facilities. This was a technological war with new types of weapons, aircraft, tanks, and ships, along with the highly sensitive development of the nuclear bomb. As a result, design and development facilities also needed to be protected from espionage, and from workers who might be stirred up by provocateurs.

The security requirements were therefore much larger and more complex than during WWI. Instead of resurrecting the USG to guard critical infrastructure a security force almost eight times larger was raised comprising of 173,000 Auxiliary Military Police and 27,000 Navy Auxiliaries. These were company employees but given special status as their work was considered essential to the war effort.

After the war in 1945, the Auxiliaries' special status ended, but most remained in post. However, the military demands reduced only briefly as the onset of the Cold War in 1947 brought about an arms race between the US and the USSR; defence production resumed and became increasingly sophisticated. Added to this, in 1950 Joseph McCarthy, a former Marine Corps officer, was elected Senator and sparked a moral panic about communist subversion. He demanded compulsory security clearances for anyone doing sensitive government work, to ensure that they could "dislodge the traitors from every place where they've been sent to do their traitorous work".

Security clearances were, and are, time consuming for the government, and sweat inducing for the applicant. The intensity of the process depends on the level of clearance required: Confidential, Secret, or Top Secret. They involve checks on the applicant's bank account, employment history, passport, driving licence, medical records, previous addresses, travel history, computer and social media use, and educational records. Often an applicant, especially for higher levels of clearance, must have a medical check-up and a drugs test. They are also interviewed about their lifestyle, alcohol and drug use, and sexual history. Their friends, employers and teachers are also interviewed about the applicant's behaviour, associations, habits, political affiliations, money management, and character. Some US agencies even use a polygraph to check physiological reactions to questions. It's a nerve tingling process usually conducted with emotional austerity by a retired cop. It costs thousands of dollars, and you can grow a beard waiting for the result. All that time you think about your past transgressions and if he believed you when you said you didn't inhale.

As well as personal security clearances, the US Department of Defence prescribed elaborate standards of security at facilities that designed and manufactured military equipment. The costs of the extensive clearances and measures were then built into government contracts. So the government, rather than the market, specified standards that it then paid for, and the wartime security infrastructure was retained, then essentially privatised, and then enhanced.

By 1950 the US Government estimated that there were 282,000 guards and private police across the country of which the majority were working on government related contracts. Forbes, the business magazine, estimated that, taking into account, informal and part-time contracts, the size of the private security sector was closer to a half million. It was clear that the war had been a massive boost for the security business.

To help companies navigate the complex government security standards, industry leaders created a professional association. Established in 1955 and known as the *American Society for Industrial Security*, it published guidance, standards, and industry news. It developed an educational framework for security professionals of which its Certified Protection Professional qualification is considered by many to be the gold standard. By 2002 in recognition of its global membership it changed its name to ASIS International and it has now grown into the largest professional security association with more than 35,000 members world-wide.

The US generally does things bigger, and often better, than anyone else, and security is no exception. In the US, a vast underpopulated continent, security was a well-established commercial service by 1900 and a multi-million-dollar business by 1950, in the UK, an ancient, settled island, it remained essentially a cottage industry until the 1980s. This reflected the major differences between the two countries. The US was a settler enterprise that drew in tens of millions of immigrants from different cultures and was not fully pacified until the late 19th century (the last major battle, known as the Battle of Wounded Knee ended with the killing of 300 Lakota People by the US Army). The UK was largely pacified by the 14th century and until the 20th century its population was mostly homogeneous and socially cohesive.

The major drivers for the early maturity of the US commercial security were the demand for secure transport services for settlers heading west, the need to control huge and restive migrant workers in the industrial era, and the requirement to guard the US's military industrial complex through the World Wars and beyond. In the UK, in contrast, the first two were absent, and its defence establishment was much smaller, it mostly used in-house guards, and the security standards were not as high as in the US. We'll see in the following chapters how security in both countries developed in the second half of the 20th century and into the 21st.

4

SECURITY FOR SALE

By 1950 the private security sector in the USA was well established although largely unseen unless you visited a defence contractor, or a major industrial site. In the UK the commercial security sector remained nascent with a handful of small security firms providing night watchmen for wealthy householders, or front-of-house security officers for City businesses. Gradually, on both sides of the Atlantic, private security officers became visible in supermarkets, nightclubs, rail stations, high streets, schools, offices, sports stadia, hotels, and shopping malls. By 2000 security officers outnumbered police officers two to one. How did security emerge from the shadows and become a feature of everyday life?

The second half of the 20th century was a period of accelerating change. The economy boomed and with it came rampant consumerism. Miraculous technology including telephones, televisions, and cars, became commonplace, and life expectancy rose by 10 years in the five decades following the war. By any material measure, life in the West had improved dramatically, yet security concerns nagged away at us, and we found that the more security we had, the more we wanted.

The Great Crime Wave

Many security concerns were linked to the crime wave that surged through the West between the 1960s and the 1990s. During this period in the US, burglaries doubled, thefts tripled, and robbery with violence quadrupled.

Criminologists pointed to multiple factors behind these trends. The population grew, and a higher proportion of people lived in urban areas where crime was more prevalent. Mass car ownership opened a new arena for theft and provided criminals with opportunities to range far from their homes. And the US and UK divorce rates more than doubled between 1960 and 1980, with the breakdown in family cohesion resulting in more delinquency.

Mass consumerism meant that many homes had the latest radio, record player, television, video recorder, microwave, and fridge. Consumerism was built on credit, because few people could afford to buy labour-saving gadgets and home entertainment systems outright. In the UK in 1970, a 22-inch colour TV cost £290, or £3,300 in 2020's money. They were so expensive that most people rented them, bought them second-hand, or took out hire purchase agreements—the "never-never" as they were known. This was a perfect economic environment for criminals. They would steal appliances from homes to sell at second-hand stores, or at the growing number of car boot sales which were packed with people who would pay cash for a bargain without demanding a receipt.

As well as homes, businesses proved increasingly vulnerable to criminals. The shift from small family-run shops to supermarkets and department stores meant that items were on open display and subject to less scrutiny. "Respectable" thieves might be inhibited from stealing from their local Mom and Pop store, but they regarded stealing from big corporations as a victimless crime. Burglary demanded a degree of skill, but shoplifting was easy. All you had to do was put an item in your pocket and walk out. In 18th-century Britain shoplifters stealing goods worth more than five shillings (about £25 in today's money) would be publicly hanged. By the 1960s shoplifters were rarely prosecuted and they were stealing goods worth around 2% of all retail revenue.

A major factor in the increase in criminality, and broader insecurity was drugs. The anti-Vietnam War movement of the 1960s with its rock music, long hair, and blue jeans is often cited as marking the start of widespread drug use. Through the 1970s, 80s, and 90s the use of harder drugs, such as cocaine and heroin, reached epidemic proportions in some cities. In the UK as much as half of all acquisitive crime was blamed on drug users, and there was a clear link between gangs that supplied drugs and upticks in violent and organised crime.

Policing the Fleecing

Until the 1960s policing was relatively straightforward. Levels of crime were manageable, there was some respect for authority, and the police enjoyed significant levels of public support. Then things became much more challenging. Rising crime undermined their reputation, and a new wave of political demonstrations and industrial disputes spilt onto the streets and the police were called upon to confront angry protestors. Racial discrimination within law enforcement, especially in the US, was also seen as a significant problem, and police brutality was captured on camera.

The first response was to expand police capacity. Between 1960 and 2000 their numbers almost doubled. In the US the number of officers jumped from 510,000 to 976,000; and numbers in the UK rose by a similar proportion, from 72,000 to 133,000. However, just putting more officers on the streets would never address the volume, and the complexity of the issues that they now faced. New types of crime demanded new types of policing. Specialised teams were needed to deal with community relations, drugs, gang violence, fraud, weapons, and public order.

In 1990 there were 2,245 murders in New York (compared with 119 in London). It seemed scarcely distinguishable from Gotham City—the dystopian, crime-ridden, home of Batman. There was a complex cocktail of crime and gang warfare, intensified by social deprivation, and racial tensions. Yet by 2000, the murder rate had fallen to 673, and by 2018 *The Economist* listed New York as the 15th most secure city in the world. What had happened?

In the absence of a real Batman, a number of factors were attributed to the drop in crime rates: more people were sent to prison; leaded petrol (which some linked to raised levels of aggression) was withdrawn; a controversial study by US academics Donohue and Levitt suggested that legalised abortion reduced the number of potentially neglected children who might later turn to crime; new video games kept young trouble makers off the streets; hard drug use declined as users were increasingly seen as losers rather than hipsters; and cheaper electrical appliances ruined the market for stolen goods. But smarter policing had an impact too.

In the 1990s the New York Transit Police developed a data-driven approach to crime reduction. The architect of the systems was Jack Maple, a stout, sharp-dressed police officer, who was fond of wearing a three-piece suit, with a bowtie, two-tone spats, and a Homburg hat. While patrolling the subways, *caves* as Maple called them, which were notorious for their violent robberies, he grew frustrated at the rising rate of crime despite energetic policing.

Maple started plotting incidents on a map, using coloured pins to mark crimes and graphs to show trends, so teams of officers could target hot spots. As a result, between 1990 and 1992 subway crime fell by 27%. The analytical, data-driven technique caught the attention of the Mayor of New York, Rudy Giuliani, who appointed Maple's boss, Bill Bratton, to be Commissioner of the New York Police Department. Bratton brought Maple with him, and his map and graph approach was computerised into a system known as *Compstat* (short for Compare Statistics). This was used to analyse crime across the city, allocate police resources, and share information between precincts.

Along with Compstat, Giuliani and Bratton pushed an approach to policing known as *broken windows*, or *zero tolerance*. The theory was that low-level crime and anti-social behaviour such as littering, graffiti, fare evasion, urinating in the street, and drinking in public, created an environment that encouraged more serious crimes and undermined everyone's quality of life. If low-level crime was addressed, it was believed that all levels of crime would drop. By the end of the 1990s crime rates in New York had halved and much of it was attributed to this new style of policing.

The approach wasn't without its critics. Some pointed out that Compstat and the broken windows philosophy came at a time when crime rates nationally were falling anyway, and others thought there was inherent racist bias in the approach. But its apparent success made Bratton famous. He was feted as a crime buster on the cover of Time Magazine. He went on to become Police Chief in Los Angeles, and then to serve a second term as NYPD Commissioner. In 2009 he received the CBE (Commander of the Order of the British Empire) for promoting US-UK police cooperation, and he was even considered for the post of London's Metropolitan Police Commissioner.

Giuliani's career was even more stellar. For his leadership after the 9/11 attacks, he was named Time Magazine's person of the year, and in 2002 he was given an honorary knighthood by the Queen. He was a presidential candidate in 2008, and, in addition to various business activities, he became President Trump's attorney, although not without controversy.

As a lawyer and a politician Giuliani's understood the trade-off between freedom and security. Early in his tenure as Mayor he said, "Freedom is not a concept in which people can do anything they want, be anything they can be. Freedom is about authority. Freedom is about the willingness of every single human being to cede, to lawful authority, a great deal of discretion about what you do and how you do it." Some saw this as alarmingly authoritarian and counter to the libertarian US spirit. Others believed that he articulated how the authorities should be empowered to create an environment for individual freedom. Either way, he might have been better understood in Beijing than in New York.

As for Jack Maple, he left the police in 1996 and started a crime-fighting consultancy, taking his data techniques to other cities including New Orleans, Newark, and Istanbul. He died of cancer in 2001 but in memory of his role in securing the streets of New York, a street in Queens was named after him in 2015. Bratton was there at the ceremony.

Compstat laid the foundation of the data-driven approach to policing that would become widespread. From pins in maps, to clunky computers, the techniques became more refined and the technology more advanced. Incident analysis would become digitised, using mass data from the internet to generate intelligence that provided real-time situational awareness, and ultimately, predictive analytics. As we'll see, these techniques have been adopted by private security companies for sale to commercial clients.

Security Surfs the Wave

Greater police numbers and improved tactics clearly had an impact on driving down crime, but private security also had an important role to play. It's hard to separate its influence from all the other crime-affecting factors, but if raw headcount is a measure, it was significant. Between 1950 and 2020 in the US, the number of private security officers more than doubled from 500,000 to 1.1 million. In the UK over the same period the increase was even more dramatic, from 5,000 to over 300,000. And on both sides of the Atlantic, private security officers outnumbered police officers.

In the 1970s private security matured and developed specialist services that were demanded by business but were not provided by the police. The police focussed on law enforcement for the public good, but that left plenty of space for private security to focus on crime prevention, for private clients, for a profit. During this period, private security developed into five distinct sectors: guarding, equipment, cash in transit, investigations, and alarms.

Guarding has always been the most visible part of private security and the presence of a security officer in a quasi-police uniform undoubtedly deters crime. However, the largest sector within private security is the provision of security equipment: locks, safes, lights, fencing, CCTV, and access control systems, which, in many cases has replaced security officers.

Cash in transit services, moving bags of money between businesses and banks, was a major sector. It grew rapidly from the 1960s, but it started to decline with the introduction of electronic bank transfers and cashless payments in

48 THE RISE OF SECURITY and Why We Always Want More

Security Officers are the most visible part of a multi-billion dollar industry.

the 2010s and was hit hard during the pandemic as the demand for touch-free transactions grew. In the US it was saved by marijuana. Although legal in 36 States, it remains illegal at Federal level, which means that banks won't take their business and most transactions remain cash. In 2022 the US legal marijuana business was worth $27 billion. With both the green backs and the green stuff being targeted by criminals, protecting the legal marijuana business is one of the fastest growing parts of the US private security market.

Investigations have also provided a steady stream of business since Pinkerton's day. These include reviewing internal thefts (or shrinkage as it is known in the trade), scrutinising misconduct, background checks on new hires, drug screening tests, and due diligence on potential business partners.

And finally, there is the alarm sector, which, as we'll see later in this chapter, creates a self-perpetuating business model.

The rise in crime was not the only reason for the growth in private security services. In the second half of the 20th century there was a huge rise in the number of private spaces that were open to the public including shopping centres, airports, sports stadia, leisure complexes, concert halls, nightclubs, and car parks. During this period there was also a decline in informal surveillance:

janitors, caretakers, servants, housewives, and bus conductors. In part, they were replaced by CCTV (described in some detail in chapter 6), which did not provide the same level of deterrence, or by contract staff, who lacked the same proprietorial attitude.

As crime rose, police resources became overstretched, and they focussed their attention on their core task of catching criminals. This forced private businesses to step up to their responsibilities to protect their own property. Gradually the police withdrew their direct support for a range of tasks including protecting cash in transit (which had been common in the UK), investigating burglar alarms, patrolling airports, and controlling crowds in sports grounds. As the police pulled back, private security jumped in.

The UK's population rose from 50 million in 1950 to 58 million in 2020, so there were more people who could be criminals, victims, or buyers of security services. And there was more stuff to secure too, as per capita income almost tripled from £7,300 to £30,000 over the same period. The insurance industry also had a major stake in crime and security. People's fear of crime resulted in more policies being taken out. The insurance industry's fear of payouts led to them imposing greater security standards on clients as a precondition of cover. And the property owner's fear of loss led them to invest more in security.

As the 20th century wore on, security systems became cheaper and more widely available. In the 1950s electronic access control, CCTV, and alarms were exclusive to sensitive military installations. In the 1960s banks started to use them, then government offices. By the 1980s they became common in upmarket stores, followed by the homes of the rich people who shopped in them. By the 1990s CCTV was in every shopping mall and high street, and from the 2000s they were within the budget of most homes.

Security was a deterrent and mark of status, both reassuring and aspirational. Yet it was also deeply unsatisfying. As Mike Davis put it in his 1990 book City of Quartz, "security generates its own paranoid demand". Its presence reminds you to be concerned, which makes you seek more reassurance. Your security is effective when you do not experience any incidents. You have security when you are not robbed or attacked. You are buying an absence of something. You are attempting to buy peace of mind. But if there is an incident, it is a trigger for yet more security measures. And if there isn't an incident it doesn't mean that you can relax your guard, it means that you might need more. The more you have, the more you want. Security is a self-licking lollipop.

You can't have too many CCTV cameras.

Segre-Gated Communities

It is sometimes said that in the first third of your life you seek sex, in the next you seek status, and in the final part you seek security. It is therefore understandable that retirees seek security within gated communities: controlled environments with high walls, access control, guards, alarms, panic buttons, and CCTV. By definition they are exclusive, impermeable to non-residents, segregating the haves from the have-nots. This is Maslow's hierarchy of needs played out in real estate. Once people have valuable possessions, they need a secure place to enjoy them, and social interaction with a community of people in the same economic bracket.

Chapter 4: Security for Sale

A segregated community in South Africa.

US sociologist William H Whyte noted that people were drawn to gated communities because of a perceived fear of crime rather than actual crime rates. Surveys showed that Milwaukee suburbanites were just as worried about violent crime as residents of inner-city Washington despite a 20-fold difference in relative levels of crime. Much of this is because, once inside a gated community, the outside world starts to feel increasingly chaotic, forbidding, and insecure. The footprint of residents' lives shrinks to the controlled, anxiety-free environments of their own community and their local shopping mall. Although with the rise of home deliveries, they can now have everything they needed dispatched to their door, and never have to navigate the menacing world beyond their walls.

From a standing start in the 1960s, the number of gated communities in the US grew to about 20,0000 by the year 2000 containing a total of around four million residents. The phenomena spread to the UK where numbers grew to 1,000 gated communities over the same period. It is easy to disapprove of these exclusive communities in environments where the fear of crime greatly exceeds the reality. But consider the case of South Africa. By 2000 around 5% of homes were in 6,000 gated communities with extensive security features and armed guards. In many ways, these represent the ultimate in segregation: wealthy and mostly white folks in fortified compounds surrounded by poorer black neighbours. But with South Africa's high rates of violent crime, they also represent a survival strategy.

Do Be Alarmed

The provision of intruder alarms is a major sector within the private security business. Like manned guarding, it received a big boost from the US government. Bank robberies may have been the stuff of western legend, but in reality they were rare until the 1920s and 30s when organised crime started to take an interest. In response, the newly formed FBI set up its *Public Enemy No. 1* campaign, aimed at encouraging the capture of high-profile bank robbers including John Dillinger, Baby Faced Nelson, and Bonnie and Clyde. As a result of the FBI's efforts, and with many potential criminals being called up during WWII, bank robberies decreased in the 1940s.

After the war, the criminal aristocracy again targeted banks, and in response the government passed the 1968 Bank Protection Act which aimed, "to discourage robberies, burglaries, and larcenies, and to assist in the identification and apprehension of persons who commit such acts." This obliged banks to incorporate a series of security measures including vaults, lights, locks, and intruder detection alarms. It was another payday for private security.

A big winner was the New York based American District Telegraph, or ADT as it became known. An early pioneer of communications, it went into the alarm business in 1874 after one of its customers was burgled and needed to call for help. ADT developed what was known as *Central Station Protective Services*. This was an intruder detector system linked to a control room that would call out the police or a private response team. By the 1960s ADT had cornered 80% of the US alarm market and was cut down to size by antitrust legislators. Despite having its wings clipped, like many of the 19th-century early starters in the security business, it continued to do well. It expanded globally and it now has around 20,000 staff and annual revenues of more than $5 billion.

In the late 1960s US forces in Vietnam started to draw down, and the defence industry looked around for new markets and found that military surveillance technology including lasers, infra-red, ultrasound and microwaves, had applications in commercial intruder alarms. Old style alarms were set off when an intruder broke an electrical circuit attached to a door or window. These new ones created an invisible fence that could identify an intruder as they approached a property. At around this time, the first Star Trek TV series was being broadcast. People became fascinated with advanced technology, creating a receptive market for new security products. But there was also a compelling economic case for them.

Business owners always balance the cost of potential losses against the cost of security. Before the crime wave, most businesses could be locked up overnight without fear of burglary. If there was especially valuable stock, a night watchman might be given a few coins to keep an eye on the property as he rubbed his hands over a coal brazier.

After the war, pay rose and conditions improved, and labour, even at the cheap end of the market, became much more costly. As we saw in chapter 2, providing a 365/24/7 guard is surprisingly expensive. Unlike Ancient Egypt where it took two people would provide round the clock cover, in the modern world with a 40 hour work week, it takes four or five people working in rotation, and as they work nights, weekends, and holidays, their hourly rate is enhanced. Furthermore, for greater safety, many advocate two-person shifts overnight (the inside joke is that allows one to sleep, while the other keeps watch), so the cost of security became prohibitive for many businesses.

But by 1970 as the crime wave continued, theft had become all but inevitable for most shops. The US Department of Justice estimated that 97% of all urban retail businesses were burgled or robbed at least once a year, and this was in addition to shoplifting.

The solution lay in those new-fangled intruder detection systems. Instead of some old dude drinking coffee and smoking cigarettes all night, a security company could wire up the premises and send a response unit if there was a break-in. Electronic alarms cost a fraction of the price of a warm body, and you didn't have to sweep butts from the hallway each morning.

New-fangled intruder detection systems.

As the 1970s progressed, intruder alarms became more sophisticated and were integrated with smoke detectors, sprinklers, and fire alarms. The use of common cabling and a single control centre provided economies of scale. Security companies shifted from being a Sleepy Joe employment agencies into high technology innovators. Insurance companies, recognising the ability of integrated systems to help control losses, offered significant discounts for properties using them. It was a win, win scenario.

A Chubb alarm central control station in the 1970s.
(Courtesy of Chubb and the London Metropolitan Archive)

The intruder detection business also generated a self-perpetuating economy of its own. The vast majority of alarms—more than 95%—are not generated by intruders, but by mechanical faults, programming issues, misuse, birds, insects, rodents, shadows, changes in temperature, or voltage fluctuations, in other words: false alarms. Invisible fences created visible concerns. Alarms reduced the overall cost of guarding premises, but businesses still needed security officers to respond to the errors generated by their own systems.

Mall Cops

The concept of the mall is hardly new; marketplaces and bazaars are as old as human history. But the modern mall can trace its lineage to 1959 in Kalamazoo, Michigan, where its car-free, secure, shopping nirvana, was the first of 4,500 to be built across the US. They needed private security officers to keep out troublemakers, control queues, guard stock, oversee deliveries, monitor CCTV, suppress shoplifting, and reassure staff and visitors that order was being maintained.

Malls are all about convenience and security so you can focus on your shopping rather than worry about traffic, the weather, or your wallet being stolen. You park your car and step into a clean, climate-controlled, brightly lit, commercial wonderland where it never rains, and everyone is happy to see you. The piped music is not just for entertainment, it's a security feature that manages emotions. The Eagles rather than Eminem set the tone, lowering heart rates, and making more shoppers docile and relaxed.

These spaces are like walled cities of the past, the owners are like feudal lords providing protection through their own security force. Inside you are cocooned and liberated from anxiety. Once you have experienced that carefree feeling when shopping, unconsciously it becomes the standard that you aspire to at work, at the cinema, at the sports ground, at your child's school, and at home. You want everywhere to feel that controlled and secure.

The Security Business

If an organisation wants security, it can either contract a private security company or employ its own in-house team. The approach taken by business is evenly split between the two. There are advantages and disadvantages with each.

Contracting private security companies is often cheaper as they can exploit economies of scale when recruiting, training, equipping, and administering staff. Their staff are not unionised, they don't acquire company worker rights, and they have fewer fringe benefits. A private security company can also offer a surge capacity to cover busy periods or sick leave, and they often have a range of specialist skills such as investigators, dog patrols, event staff, and counter surveillance teams, that can be deployed on demand.

Advocates of in-house security cite the positives: loyalty, knowledge of sites and procedures, lower turnover, understanding of the company culture, and the prestige of having their own team. If you ask a security officer which they would prefer, without hesitation, they will opt to be in-house as they tend to be better paid, treated with more respect, have greater job security, and they enjoy being part of the company "family". But for the employer there are also negatives: in-house teams are often more expensive, it can be harder to let go of under-performing staff, and their professional expertise can be quite narrow.

Businesses often turn to hybrid solutions with a combination of in-house managers and contracted security officers. This utilises the advantages of both approaches: in-house company knowledge and loyalty, and the cost-effectiveness and flexibility of contractors.

Whether in-house or on contract, few people grow up aspiring to be a security guard. The job attracts those with limited opportunities, immigrants, or people who have fallen on hard times. Those for whom the alternative is driving a taxi or flipping burgers. Most have little by way of academic qualifications but managed to steer clear of trouble and keep a clean police record. The job never brought prestige or financial rewards, turnover rates are high, the work is normally dull, the hours are unsociable, and the conditions are often poor.

Who aspires to be a security guard?

Milton Lipson's classic 1975 book *On Guard* quotes Edward Lee, a city official in Philadelphia discussing the licencing of private security guards, "They are paid minimal wages to take physical and verbal abuse, expel troublemakers, catch bullets during a robbery, outsmart shoplifters, protect valuables, and save store personnel from the ravages of roving gangs. These are Herculean tasks entrusted to people who are, for the most part, under educated, poorly trained and poorly supervised.... [yet] they are just as vital as city police in the day-to-day operations of many local retail businesses, factories and schools." Five decades later his points remain pertinent.

In the US being a security guard is one of the highest-risk occupations. The US Bureau of Labour Statistics show that in 2003, for every 100,000 workers of all types in the US, four were killed at work. But security guards had over double the fatality rate with more than nine dead per 100,000, of which 60% were as a result of acts of violence.

But they could dish it out too. Milton further quoted Edward Lee, "these guards are armed with revolvers, mace, tear gas, knives, attack dogs, billy clubs, and occasionally 12 gauge shot guns." Lee added, "they lack the authority to arrest, but they have the power of life and death on their hips. It is a fact that security guards in this area shot 13 people." In the UK, where guns are illegal and only 5% of the police are armed, it comes as a surprise that security guards in the US routinely carry lethal weapons and the security profession is often referred to as "gates, guards and guns."

Despite the risks involved, the security sector was, and continues to be, very lightly regulated. In the US, only 38 of the 50 states have licencing requirements for security officers. These might typically include an age limit of 18 or 21, a criminal record and firearms certificate check, general security training of between 16 and 40 hours, and firearms training of between 10 and 40 hours. As Mike Davis put it in City of Quartz, "It is easier to become an armed guard than it is to become a barber, hairdresser or journeyman carpenter."

In the UK the regulation of the security sector is also light and is aimed at protecting the public as much as improving standards. The first law introduced was the 1975 Guard Dog Act which prohibited the use of dogs unless they were under the control of a handler. It followed the killing of three boys by guard dogs in two separate incidents in the early 1970s.

The second law aimed to address the more widespread problem of pub and club door supervisors, better known as *bouncers*, who were often more renowned for their boxing than their customer service skills. Clashes between alcohol-fuelled customers and testosterone-fuelled bouncers frequently led to violence, sometimes extreme, and occasionally fatal. In response, the government introduced the Private Security Industry Act of 2001 which aimed to raise standards through monitoring, licencing, and 32 hours of compulsory training for front-line security staff. For the first time in the UK, private security was held to account. Standards undoubtedly improved, and bouncers learned to de-escalate rather than enflame difficult situations.

Bouncers became licenced Door Supervisors, but they kept the same hairstyles.

I had reason to welcome the Private Security Industry Act. In the late 1980s, I was violently set upon by bouncers in a London club and flung out the door. I don't know what I had done to attract their attention, but as I arced through the night air, just before I landed painfully on the concrete steps, I thought to myself, "these chaps could benefit from some customer service training."

Growth and Consolidation

Across the US by the 1960s there were thousands of private security companies responding to the growing demand. While some expanded by increasing their range of services, winning new business, and by purchasing existing security companies, the sector remained dominated by a handful of early pioneers.

Pinkerton was the biggest provider of guards but it avoided providing services for labour disputes following the fatal events at the Homestead Steelworks (described in chapter 3). Its main rival, Burns, developed a national presence in the 1950s. Brinks expanded its cash in transit business and by 1969 had 1,100 armoured vehicles. The alarm sector was dominated by ADT, while the new kid on the block was Wackenhut, which was founded by a former FBI agent with an eye for government contracts, by the 1970s it was the US's fifth largest security company.

Over in Sweden, in the 1930s, Erik Philip-Sörensen founded Securitas. In 1981 the company was split between his sons Jörgen and Sven. Jörgen headed the growing British operation which he named Group 4 after merging four separate companies. Sven took over Securitas Sweden, but, to the dismay of his family, he soon sold it and the brothers never spoke to each other again. Under new corporate ownership Securitas expanded rapidly by acquiring other companies including Pinkerton and Burns. By 2000 it was one of the largest security companies in the world.

Jorgen obviously had a point to prove. In 2000 he bought Falck, a Danish security company, followed in 2002 by Wackenhut, the US company which gave it access to US government business. In 2004 he bought Securicor, one of the biggest British security companies. Securicor had started out in 1935 as Night Watch Services with a dozen guards wearing old police uniforms and patrolling wealthy London residential streets. It paused operations during the war and was rebranded The Security Corps in 1946 as it switched focus to industrial security services. It softened its military image by changing its name to Securicor in 1953 and developed a wide range of security services including cash in transit and prisoner transfers.

With the acquisition of Securicor, Group 4 was rebranded as G4S, overtaking Securitas to become the world's largest security company with more than half a million staff and revenues of close to £6 billion by the time of Jorgen's death in 2010. But the dog-eat-dog story doesn't end there. In 2020 the US security and facilities management company, Allied Universal, announced a takeover of G4S. The combined company with revenues of $18 billion and 800,000 staff became one of the world's largest employers.

Since the end of WWII the demand for private security has grown steadily. By 2022, the US market was worth $52 billion and the UK market was worth $12 billion. On both sides of the Atlantic there were twice as many private security officers as police officers. The crime wave, the growing number of private venues, and the introduction of new security technology, had been good for business. But, as we'll see in the next chapter, there was another factor too.

5

GENERATING ANXIETY

By the late 20th century people in the West were living longer, healthier, safer, and more secure lives, than any previous generation, yet, paradoxically, the fear of loss, injury, and premature death grew to fetishistic proportions. The private security industry may have used stoke and soothe methods to market their services, but it alone didn't generate anxiety. Much broader and more powerful forces were at work.

Growing up in England the 1960s and 70s, my childhood was fairly typical. Playing in the streets, exploring the woods, climbing trees, fishing, playing conkers, riding homemade carts, and building bonfires. Like most of my generation, I walked to school and was allowed to roam free so long as I was home before the streetlights came on. The closest I came to parental supervision was when my mum would say, as I lit another firework, "it's all fun and games until someone loses an eye," or when she gave me 20 pence and asked me to walk the half a mile to the shops to buy her a packet of cigarettes. Our family's Austin A40 hatchback had no seat belts, and my favourite place was, with the dog, in the boot, facing backwards, our breath condensing on the rear window.

I think I had a brilliant childhood, but if a child engaged in those activities in the 2020s, they would be taken into care and their parents would be charged with neglect. It's clear that society's attitudes shifted dramatically over the course of half a century. German academic Ulrich Beck described our world as a *risk society*, "increasingly occupied with debating, preventing and managing risks that itself has produced." It is, he said, "an inescapable structural condition of advanced industrialization". It seems that as we become wealthier, we become more risk aware, and we demand more safety and security.

The focus of this book is security, which is about protection from people, rather than safety, which is about protection from things. The distinction is important to those working in the business, but many people use the terms interchangeably, and often they are bound together, as in "safety and security."

The amygdala, the part of the brain that reacts to fear, does not distinguish between safety and security. If we are about to be punched in the face (a security issue), or if our car is about to crash (a safety issue), the emotional and physiological reactions are identical. Both generate fear, our heart rate explodes and, gasp—we take a sharp breath—as we oxygenate our blood in preparation for fight or flight.

Assuming we survive the punch or the crash, the amygdala also plays a deeper role in shaping our behaviour. Jon T. Willie, neurosurgeon at Emory University in Atlanta says, "if you have an emotional experience, the amygdala seems to tag that memory in such a way so that it is better remembered." In other words, we heed the old proverb, *once bitten, twice shy*. And we don't just react to being bitten ourselves; if we see someone being mauled by a dog, we steer clear of dogs in the future. If we see a horrific car crash, we will instinctively moderate our own driving. As the 20th century wore on, an increasing range of new things including flying, nuclear war, terrorists, air pollution, AIDS and a potential zombie apocalypse, stimulated our amygdala and filled us with fearful emotions.

Our savannah-adapted physiology is not always helpful when navigating the modern world and neither is our language. We often pose the question, "is it safe?" Safe is an absolute term that implies something is either safe or it's not, and it demands a binary answer: yes or no. The problem is that very few things are absolutely safe. Even our homes, where we should feel safest, have stairs, knives, candles, electricity, and boiling water that can harm us. Crossing the road, taking a train, or playing football, are not safe. Our language forces us to make a proposition that we cannot fulfil.

If we want to be more constructive, we should ask "what are the risks?" This calls for a considered, Dr Spock-type response, that explains the likelihood and the impact of something that might cause harm. But we are not all risk managers or Vulcans. We are humans with amygdalas and imprecise language. Neither are efficient at evaluating and describing modern safety and security issues, so both contribute to unrealistic expectations.

Our preoccupation with safety has entered our everyday language. Google's ngram facility that charts the frequency of usage of words and phrases, shows that the phrase "take care" was common in the US during the world wars, which given the gravity of the situation, was a reasonable suggestion. After declining in usage for a few decades, the phrase became common again in the 1980s when it was often used instead of "goodbye."

Even stronger was the use of the phrase "stay safe". According to ngram, this appears to have entered our lexicon around 1990 and rapidly became a cool and caring way of saying goodbye. It carried the implicit message that danger lurked, and you needed to be vigilant. The origins of the phase are obscure, but in the 1990s, Paul Moxness, Head of Security at the Radisson Hotel Group and prominent security practitioner who regularly presented at international events, started using "Stay Safe" as an email sign-off. It could be that his catchphrase may have rubbed off onto the wider security community and from there it seeped across to the public who had a ready appetite for caution.

Seeing is Believing

Most people throughout history could not read or write. Until relatively recently people received news through that most unreliable of means—by word of mouth. It was only from the 1850s that the UK's literacy rate exceeded 50%, fuelling a boom in mass-market newspaper readership. Newspapers provided a wider and more consistent dissemination of news. Reading, however, is a learnt rather than an instinctive means of receiving information. The written word certainly engages our intellect, but it doesn't fire our emotions with the power of the spoken word. So the arrival of the radio broadcast in the 1920s, with the authoritative tones of plummy-voiced BBC presenters, quickly found a receptive audience. Radio made people much more emotionally involved with what was going on in the world. And alert to its hazards.

But whatever newspapers and radio could do, television could do better. Emotions could be supercharged by viewing moving images. Seeing is the most natural, immediate, and intense means of absorbing information. Seeing is believing, television takes you there. The BBC made its first televised broadcast in 1936, but suspended services during the war because there was concern that German bombers could use TV signals to identify targets and because it was considered too powerful a medium for a country in conflict. Battlefield images might undermine morale and support for the war.

Television broadcasts resumed in 1946 and by 1960 almost half of all UK households had a TV. They were so expensive that most people rented rather than bought their sets. But it was worth it; everyone was transfixed by moving images. Soon there was concern about how long people were spending in front of the TV. It was an irresistible *magic box*, that provided a front-row seat at a coronation or royal wedding, it offered a view from the terraces of football grounds, and it allowed people to accompany astronauts on their first moon walk, all from their own sitting rooms.

It is easy to forget the novelty and excitement of the early years of television. At that time most people had grown up in homes with coal fires and outdoor toilets, where entertainment was a chat around the kitchen stove, or reading a book by flickering lamplight. The masses did not get electricity in their homes until the 1930s. If anyone had described a TV to people a century earlier, they would have been locked up as lunatics, two centuries earlier and they would be hanged as heretics.

But once we had it, TV grabbed our attention and manipulated our emotions, including our fears, like nothing before. In July 1966 more than 30 million people were enraptured by the live coverage of England's victory in the football World Cup. In October of the same year, not long after the beer bottles and the bunting had been cleared away, everyone tuned in to learn about an appalling tragedy at the Welsh mining village of Aberfan. A mountainous coal slag heap had collapsed onto a school killing 116 children and 28 adults. The nation was stunned.

Terrible things had happened before: war, plagues, shipwrecks, fires, and, just 25 years earlier, the blitz which had killed tens of thousands in the UK and left millions homeless. But the television images of Aberfan brought a level of emotional engagement never experienced before. People witnessed the disaster, the rescue crews at work, the tortured faces of the bereaved, and the mourners weeping over tiny coffins. Television took viewers to the scene and showed the devastation and the grief up close. It was both compelling and disturbing. As journalist Andrew Marr put it, "Hard news really is hard. It sticks not in the craw, but in the mind. It has an almost physical effect causing fear.... or shock".

In 1955 an independent TV channel—ITV—was launched to provide competition to the BBC's patrician coverage. Unlike the BBC, which was funded by licence fees, ITV raised its revenue through advertising which meant that it hunted for emotional engagement to boost viewing figures. Newspaper editors had long understood the public's fascination with sensational death, and television built on the mantra, *if it bleeds it leads*. Fear-based media has continued to manipulate our emotions ever since.

The second half of the 20th century provided no shortage of material in the UK alone: the Hither Green Rail crash of 1967 left 49 dead, the Ibrox football disaster of 1971 left 66 dead, the Staines plane crash of 1972 left 118 dead, the Moorgate tube crash of 1975 left 49 dead, the Manchester air disaster of 1985 left 55 dead, the Zeebrugge ferry disaster of 1987 left 193 dead, the Piper Alpha oil rig disaster of 1988 left 167 dead, the Kegworth air crash 1989 left 47 dead, the Marchioness sinking on the Thames of 1989 left 51 dead. On top of this the *Troubles* in Northern Ireland left 3,500 dead between 1969 and 2000, with bombings, riots and murders reported on TV almost nightly.

The point of news is to report the exceptional. This is why the 20 people dying every day in car crashes for the best part of a century isn't a story, yet one person dying in a terrorist attack hits the headlines. The result is that our perception is distorted. As US academic Michael Shermer wrote, "The idea that we live in an exceptionally violent time is an illusion created by the media's relentless coverage of violence, coupled with our brain's evolved propensity to notice and remember recent and emotionally salient events." All this causes us to react disproportionally to the extraordinary.

The ultimate TV spectacular was the 9/11 attack of 2001. The images of the planes flying into the twin towers, people jumping from windows to escape the ensuing inferno, and the buildings collapsing, are indelibly seared into our consciousness. Osama bin Laden knew exactly what impact the attacks would have, gloating shortly afterwards, "America is full of fear from its North to its South, from its East to its West. Thank God for that."

Had 9/11 occurred just a few years later after the introduction of social media (Facebook was launched in 2004, soon followed by YouTube, Twitter and Instagram) the attack would have had even more impact. Social media quickly proved popular, and were turbocharged by the arrival of the iPhone in 2007 which took it from a desktop computer at home, to an app in our pockets. Everyone was suddenly able to post footage of incidents and their reactions to them. Everyone was at the centre of their own news story, broadcasting their own experience of events. What the personal reports lacked in analysis they made up for in emotion. If you were unable to articulate your feelings in words you could find an emoticon that did it for you. Social media provided a level of immediacy and intimacy that traditional channels could not match. And it spread misinformation and fake news, that could go viral and further fan the flames of anxiety.

"I've looked a lot at why we are more fearful now than 200 years ago," says Margee Kerr, a sociologist at the University of Pittsburgh. "And one thing that keeps coming up is the immediacy with which we get the news. This makes it feel more emotionally charged. We start receiving notifications on our phone as soon as these disasters happen. So there's a false sense of involvement that we didn't have years ago."

As if there wasn't enough death in the media, film-makers cashed in on our morbid fascination with a series of movies about real or imagined disasters: The Day the Earth Caught Fire of 1961, Krakatoa East of Java of 1969, The Poseidon Adventure of 1972, The Towering Inferno of 1974, Meteor of 1979, Virus of 1980 and Alive of 1993. These movies contributed to the *Jaws effect*: where situations depicted in movies become real-world anxieties. If you have seen the 1974 movie Jaws you've probably been nervous about entering the sea ever since. Even years later the soundtrack: *bom bom...... bom bom...... bom bom...... bom bom bom bom bom bom*, plays through your mind as you swim in the blue.

There was clearly a market for the manufactured feelings of fear and anxiety. It turned out that people liked to join others in comfortable movie theatres and share base emotions, in total safety.

Alison Holman of the University of California was studying mental health when the Boston Marathon was bombed in 2013. Three people were killed, and dozens were wounded in the attack. Holman's research had surprising conclusions. She found that those who watched more than 6 hours a day of TV coverage of the event in the ensuing days experienced higher levels of acute stress than those who witnessed the events on the ground. This was proof of the power of TV to supercharge emotions.

Psychologist Dr Deborah Serani highlighted the consequences of fear-based media including general feelings of insecurity, overestimating the odds of becoming a victim, the belief that crime rates are rising, and that the world is getting more dangerous. Writing in Psychology Today in June 2011 she explained that "fear-based news stories prey on the anxieties we all have and then hold us hostage". There is little doubt that vivid images on TV, film, and the internet, amplify our concerns about our personal safety and fed into a desire for yet more security products and services.

Industrial Safety

Moving images had a huge impact on public fear, but the origins of the safety agenda can be traced to the gunk and grime of the early 19th century. During the industrial revolution people moved from the countryside into towns, drawn by the prospect of regular work in the factories which were experiments in technology, organised labour, and economics.

That hard work and low pay were features of factory work came as no surprise: it had been much the same for agricultural work since the beginning of time. But the *dark satanic mills* of the new industrial era were an entirely new environment. Horsepower was swapped for steam power, there was scant awareness of the danger posed by machines, many industrial materials were hazardous, and there was little concept of safety. Furthermore, in the absence of clean drinking water, people drank ale or cider. This may have warded off some sickness, but it meant that people worked while at least partially drunk. Unsurprisingly, many workers were killed or injured by machines, or died early as a result of industrial diseases.

The perils of early industrialisation did not only affect adults. Until 1880 there was no compulsory school. Education was limited to the elite and to the few able to access church-sponsored schools. Throughout history children had worked the land with their parents almost as soon as they were able to walk. It was only natural that they too would join their parents and work in factories or mines. We now think that child labour was demonic, but in those times it was an arrangement that suited everyone: factory owners benefited from nimble fingers and cheap labour, parents were able to supervise their children whilst they worked in an era before childcare, Lego, or daytime TV, and children learnt a trade, earned a wage, and secured a position for the future. The downside was that playful children and powerful machines were a dangerous combination that resulted in many young deaths.

Children being mangled by machines caught the attention of early social reformers who promoted a series of measures to improve factory conditions. The first was the Factory Act of 1802 which aimed to improve workplaces for young apprentices. It obliged employers to clean their factories with limescale at least twice a year, ensure adequate ventilation, limit working hours to 12 hours a day, and offer basic education. These seem strikingly modest provisions, but this was at a time when boys as young as 12 served on Royal Navy warships, and when the masses did not have access to soap for another 50 years.

Additional legislation during the 19th century mandated inspections, further reductions in working hours, protective shields on machines likely to cause injury, and, from 1880, allowed workers to claim compensation if they were injured at work.

There is little data available to help us estimate the overall scale of industrial safety issues, but figures from the highest-risk occupation, coal mining, provide an impression. In Barnsley, in 1866, an explosion at the Oaks Colliery killed at least 360 miners, and in Glamorgan in 1894, at the Albion Colliery, an explosion killed 295. By the 20th century, safety legislation started to reduce casualties, but even so, as late as 1947, across the UK, 618 miners were killed on the job.

These figures were insignificant compared with the hundreds of thousands killed in the world wars. The prevailing attitude until the late 20th century was that industrial accidents were modest in number, unfortunate in nature, and an inevitable result of industrialisation. Although some measures had been taken to improve safety, it was not until memories of the wars had faded, the power of Trades Unions had grown, and the US had led the way with its comprehensive Occupational Safety and Health Act in 1970, that the conditions were ripe in the UK for a step change in its approach.

This came in the form of the creation of the Health and Safety Executive (HSE) in 1974. The HSE, as the name suggests, made employers responsible for health, safety, and welfare at workplaces, introduced the compulsory reporting of dangerous occurrences, and beefed-up inspection regimes. There was a new notion that accidents were not inevitable, but avoidable. Suddenly employers were forced to take responsibility for blood spilt on their factory floors.

The application of HSE directives was uneven in its early years, and old-fashioned attitudes lingered in many areas, as I discovered the hard way. My first job after leaving school was working 12 hours a day as a deckhand on a mud-scooping dredger on the South coast. One day a clumsily manoeuvred tugboat hit the dredger whilst I was climbing an external ladder. I was pitched headfirst into stinking, foaming, water, goop sliding from my hair as I emerged from the waves. I swam to a nearby wharf, jeered at by my shipmates, and clambered ashore where I was told to get cleaned up. I rode the 10 miles home on my motorcycle, took a bath and changed my clothes. Returning to work later the same afternoon I received only a stern nod in acknowledgement from the captain. At the end of the week, my wages were £3 short as I'd been docked 3 hours pay for my absence. I occasionally reflect on how the incident might be handled today.

In its first year, the HSE recorded 651 work related fatalities. By 2019 the figure had fallen to 111, so it was certainly having an impact (helped by a transition from heavy to service industries, so more people were working in offices than factories). Yet the HSE was not without its critics and the *Elf and Safety* culture was ridiculed. The premise was that the traditional British way of life was being eroded by mindless bureaucracy. There were stories about candy floss being banned in case people tripped and impaled themselves on the stick, or graduates being told not to throw mortar boards in the air in case they hit someone in the eye. The press loved these stories because they outraged people, leading to higher newspaper sales. The new safety culture was a gift to martinets, a particular breed that loves to enforce rules, real or imagined. From being a fringe issue, safety was now mainstream, and it was on everyone's mind.

As well as saving lives, health and safety also changed people's wardrobes. In 1992 the UK government introduced regulations on personal protective equipment (PPE) at work which obliged employers to ensure that workers at risk had hard hats, safety glasses, appropriate footwear, and so on. In places where there was no specific hazard, safety-conscious employers provided high-visibility yellow vests. These quickly proliferated amongst drivers, builders, maintenance workers, cleaners, and even children on school trips.

"It's yellow, it's ugly, it doesn't go with anything, but it could save your life."

Safety became something that you could display, a virtue that you could project. Yellow vests legitimised the wearer and gave them a degree of authority. They were even promoted by fashion designer Karl Largerfeld in a French information campaign saying, "It's yellow, it's ugly, it doesn't go with anything, but it could save your life." In England, the national costume was often considered to be a suit anc a bowler hat, but by 2000 it was surely displaced by the yellow vest and the hard hat. Safety has certainly moved close to the centre of our national culture.

*i*ety

Had Carl Benz been told, as he made his first car in 1885, that his invention would go on to kill millions of people, would he have hesitated? Since time began, people had fallen from, or been trampled by, horses, the main form of transport until the 20th century. They had a reputation for being *dangerous in the middle and unsafe at each end*. Back in the day safety was of little concern, accidents were part of life, it was the price you paid for mobility.

The first legislation on road transport came 20 years before the car and was aimed at steam-powered traction engines whose hissing and belching scared horses, and whose weight damaged roads. Known as the Red Flag Act of 1865, it obliged someone to walk ahead of *self-propelled vehicles* carrying a red flag (a century later and it would surely have been someone wearing a yellow vest). The Act was repealed in 1896 and motorists drove the 60 miles from London to Brighton to celebrate, an event that has been repeated annually ever since.

It wasn't long before cars started to kill people. Bridget Driscoll became the first recorded road fatality when she was run over by a car as she crossed a street in Crystal Palace, London, in 1896. She was the first of many. By 2020, more than 575,000 people had been killed on the UK's roads. In the early years of motoring, casualties got little more than a glance in the rear-view mirror. But in 1916, the London Safety First Council (the forerunner of the Royal Society for the Prevention of Accidents), was created to tackle the "alarming increase in traffic accidents." There were no published road casualty statistics at the time, but the launch of Safety First in London when thousands were being killed every week on the western front, was a sign of deep concern.

The actual scale of the problem was not apparent until 1926 when road crash data was consolidated for the first time showing that 4,886 were killed on the roads that year. By 1930 annual fatalities reached over 7,000 and the government was forced to act. To bring order to the anarchic roads, the Highway Code, a set of rules for vehicles, was introduced in 1931 and in an effort to improve driver competence, tests were made compulsory in 1935.

WWII stalled progress in road safety, but it was back on the agenda when, in 1966, fatalities spiked to 8,000. This reflected greater car ownership but the major factors for them crashing were driver behaviour, principally speeding and drinking. For the first 80 years of motoring, people could drive to a party, restaurant, or pub, drink as much as they wanted and drive home as fast as they wanted, fearing a hangover rather than a fine. All that changed in 1967 when drink driving laws, and a national speed limit of 70 mph, were introduced. It took a while for the laws to have an impact but by 1980 the fatality rate, even as car ownership continued to rise, fell below 6,000 for the first time since the 1920s.

Further road safety legislation followed in 1973 when helmets for motorcyclists became compulsory and, following the *clunk click every trip* TV campaign featuring Jimmy Savile (who later transpired to be anything but a safe pair of hands), the wearing of seat belts became law in 1983.

Improved road layouts, physical barriers between vehicles and pedestrians, compulsory child seats, and a series of road safety campaigns, contributed to further declines in casualty figures. Car manufacturers were initially reluctant to emphasise safety as they felt it would draw attention to the fact that, during the 20th century, their products had killed almost as many people as two world wars combined. It was automotive engineers from Sweden's Volvo that led the car safety agenda, marking a long journey from warrior Vikings to safety pioneers.

Being on the roads was the single riskiest activity that people did in their everyday lives. As awareness of the dangers grew, people become more safety conscious and drove more carefully. Safety was no longer something industrial and intangible, it was domestic and evident. Individual actions could save lives. Safety was a good thing.

The Nanny State

Industrial and road safety were both government initiatives. They were part of a process where the state broadened its traditional role from protecting against foreign aggressors, to protecting against criminals, to protecting against health and safety risks, and finally to actively promoting well-being and longevity. This was an inevitable consequence of the formation of the welfare state the seeds of which were sown by 19th century social reformers. By the 20th century these seeds had blossomed into workers' rights, pensions, and health care, all underpinned by the belief that the government should actively look after its people.

The term *Nanny State* was minted in 1965 by the card playing, D-Day veteran and Conservative politician, Iain MacLeod, who expressed concern about a government ban on cigarette advertising on TV. It became a favoured line of libertarians as more health and safety advice was woven into law. MacLeod's views were echoed by another conservative politician, Harold

Balfour, a Sopwith Camel pilot in WWI, who said that seat belt legislation was "yet another state narrowing of individual freedom and individual responsibility". It was a fair point, but a welfare state that provided free health care had an interest in reducing the costs of treating injured drivers and smokers' self-inflicted damage.

MacLeod and Balfour were men of their times who had grown up in a world where safety had meant crouching behind a sandbag to avoid enemy fire. But there had been a quiet revolution in their lifetimes. Huge strides had been made across a wide field of safety issues including: drinking water, food hygiene, child-birth, anti-biotics, pasteurised milk, insect control, as well as at work, and on the roads. The key indicator of their collective impact was that life expectancy between 1900 and 1980 increased by a third, from 50 to 75 years. Safety had become part of the wider rights revolution that included workers' rights, human rights, women's rights, gay rights, and minority rights. People started to believe that they had a right to be safe and that the government had an obligation to deliver. Safety had become a cardinal virtue and a political issue.

Child Safety

Nowhere was the evidence of new levels of safety more evident than in how children were treated. Within the space of a century, a change in social values and legal frameworks meant that children were no longer required to long hours work in risky, unsanitary conditions. While in the 1870s children were cleaning chimneys, by the 1970s there were warnings on toys about choking hazards. Child safety became the most important thing in the world. Children became hyper-cocooned, never let out of sight of a parent, and guarded against anything life might throw at them. All of our safety concerns were projected onto our children.

Until the 80s a child would be teased if their parents drove them to school: everyone walked. Now children are strapped into the safety seat of SUVs with side impact protection and airbags. Bicycle helmets were unheard of in the 1970s. Today they are essential kit for cycling, skateboarding, and scootering, often accompanied by knee and elbow pads and a yellow vest. Children no longer roam freely; they are under constant parental supervision and if they are out of sight, you can monitor them on your smartphone. The *Family Locator* app "comes with a great mix of safety features to keep your worries at bay. Family Locator app will let you know when a family member's battery is low ... so you won't panic if your child or spouse

Chapter 5: Generating Anxiety **73**

Constant parental supervision: don't let them out of your sight!

doesn't get in touch. You can also set alerts for when members reach a certain place, and track average speeds of drivers in the family." Surveillance parenting has arrived.

In the US, often ahead of the child safety curve, an advertisement runs: "Moms always know what is important for their kids: use seatbelts in the car; wear a helmet when they ride a bike; apply sunscreen before they go outside; and all the other things that you do for your kids' health and safety." It features a mother looking at her daughter who smiles as she clutches her "Lifeplate personal bulletproof insert." The fear of school gun violence is soothed by purchasing ballistic shields for your child. How long before these become standard for all American school children?

Most would argue that improved child safety is a good thing and that it is largely driven by parents' anxiety, rather than by nanny state directive. Who wouldn't want their child's head to be intact after a fall off a bicycle? Who wouldn't want to increase their child's chance of survival is there is a classroom shooting? Who would not want to know the moment their child was lost, distressed, or abducted?

All these are entirely understandable notions, but bubble-wrapping children has consequences. A generation that has grown up under constant supervision with safety as the dominant theme in their lives, will mature into adults with

Essential kit.

an extraordinary level of risk aversion. Life today is, by any measure, much safer than ever before. Yet we control our environment ever more tightly in a quest to make everything even safer. We seek greater protection because human nature makes us strive to improve whatever we have. Safety is part of a general advance in society.

Safety and security are bedfellows. Most people would recognise only a shapeless lump under the duvet and make no distinction between the two. Most organisations have a department that manages both safety and security, and the terms are often used interchangeably. So, the quest for ever-increasing security is interwoven with the quest for ever-increasing safety. The initial trigger was awareness of the carnage in factories, down the mines, and on the roads. Then our emotions were manipulated by the media, and then we demanded levels of safety for our children that would astonish previous generations. But previous generations don't matter, it's the current generation who will shape the future, and they will inevitably seek to make their children's lives even safer and more secure.

6

THE ALL-SEEING EYE

We all behave better when we are being watched. We mind our table manners, we don't drop litter, and we don't steal. We fear ridicule, fines, and incarceration. Watching others—surveillance—has long been a way of moderating behaviour, ensuring obedience, and achieving societal control. Surveillance is key to security.

People have always lived closely together, eating, working, and sleeping communally, constantly observed by family and neighbours who see if anyone veers from social conventions. The concept of privacy is very recent. Even in the West it is only from the late 20th century that most children have stopped sharing bedrooms with siblings, while in much of the world entire families still live in a single room. Even now, privacy is a luxury available to those able to afford large homes with big gardens and high fences. Talk about privacy to most people living outside the West and they will blink at you.

The irony is that, as we gain privacy from human surveillance, we rely more on artificial surveillance to underpin our security. While many of us remain strangers to our neighbours, we share everything there is to know about ourselves with automated systems operated by faceless organisations, the consequences of which remain open to speculation.

In this chapter we look at the various stages of surveillance starting with the *imagined surveillance* promoted by religions whose omnipresent God supposedly had an all-seeing eye. In the 19th century, prison design incorporated *natural surveillance* where a small number of unseen guards could observe a large number of prisoners. Natural surveillance was incorporated into urban design in the 1970s as a crime prevention measure. From the 1980s, *technical surveillance*, in the form of the cold probing eye of closed-circuit television (CCTV) replaced the sleepy-eyed watchman and the curtain twitching old lady. By 2020 there were a billion CCTV cameras worldwide, and surveillance had become a $20 billion-a-year slice of the commercial security industry.

From imagined, to natural, to technical surveillance, the next stage in the journey is *dataveillance* which tracks you through internet activity, facial recognition technology, global positioning systems (GPS), artificial intelligence, and smartphone data. One day we may look back fondly at those few decades, between our escape from the constant gaze of our community, and the quiet arrival of ubiquitous dataveillance. A time when lipstick on your collar rather, than your search history, told a tale on you.

Imagined Surveillance

A key purpose of religion was to lay down and enforce codes of behaviour. If everyone followed them, people would live peacefully together. There would be no crime or immorality, and no security problems. A powerful element of Christianity was the concept of an omnipresent God who observed your every action and sat in judgement at the end of your days. The message was clear: if you were good, you were given a set of big white wings and went to heaven. But if you had done something bad, even if it went unnoticed by your neighbours, you didn't escape God's gaze, and you'd end up stoking the fires of hell.

He knows if you've been good or bad.
The All-Seeing Eye on the Hartebrug Church in Leiden, The Netherlands.

This concept of *imagined surveillance* was reinforced in many churches using a symbol of an all-seeing eye. This was represented as a single, expressionless orb, making silent judgement on thoughts and deeds. The symbol is still used in the national iconography of several countries including Poland, Ecuador, and Serbia, it's inscribed on the great seal of the USA and it features on the one-dollar bill, although perhaps one day it will be replaced by a CCTV camera.

Natural Surveillance

The next stage in the evolution of surveillance was the *panopticon*. This was a building where many occupants could be observed, unseen, by a handful of people in charge. It was promoted in the late 18th century by English philosopher Jeremy Bentham who believed that we should strive for, "The greatest happiness for the greatest number of people". This was the foundation of his rational approach to surveillance. Most people would accept it for the security benefits that it brought. Some might object on privacy grounds, or because they wanted to do something illegal, but the majority view should prevail.

A *panopticon* was circular or star-shaped in plan, and could be applied to hospitals, asylums, factories, schools and—classically—to prisons. At the centre was a screened hub from where all activity could be observed. This created, in Bentham's words, "the sentiment of a sort of invisible omnipresence." His panopticon went on to influence prison design around the world. The most spectacular example, built in 1926, was the *Presidio Modelo*, (Model Prison) in Cuba. Its five enormous panopticons had a honeycomb of 1,500 individual cells along the circumference of the five-storey circular building. From a central tower, a single guard could view every prisoner.

While Bentham promoted the panopticon as a rational, efficient, and benign means of maintaining order, to others it was a sinister means of control. In *1984* George Orwell captured the chilling side of surveillance saying, "there was of course no way of knowing whether you were being watched at any given moment... you had to live... in the assumption that every sound you made was overheard, and, except in darkness, every movement scrutinised." In the final chapter of this book we'll see how prescient he was.

Orwell always trumps Bentham in conversations about surveillance, his images of dark totalitarian states fester in our imagination. But natural surveillance is not just a technique for institutional control. It can also promote benign natural security. In the 19th century, some communities around the North Sea and Baltic Sea used, as they are known in Dutch, *spionnetjes*, or spy mirrors. These mirrors, about the size of your hand, were fixed outside windows and angled to allow observation of the street, left and right. From a comfortable chair in their living rooms, people could see the comings and goings of their neighbours. In the Dutch town of Leiden, many homes still have *spionnetjes*, and on the classical pediment of its Hartebrug Church, a huge motif of an all-seeing eye peers over the rooftops. With so much natural and imagined surveillance, everyone should be on their best behaviour.

As we saw in chapter 4, a crime wave spread across the Western World in the 1960s, which rekindled interest in natural surveillance. In 1971, criminologist C. Ray Jeffery published a book entitled "Crime Prevention Through Environmental Design". CPTED as it became known, built on the classical notion that people behave better when they think that they are being watched.

Jeffery believed that natural surveillance could be created by encouraging social interaction, increasing pedestrian and bicycle traffic flows, having windows and balconies overlook public spaces, landscaping open areas to create a pleasing environment, and by using lighting to reduce the fear of walking at night. He also recommended natural access control by limiting points of entry, facilitating observation by keeping walls and vegetation low, and designing out easy access to windows and rooftops by intruders.

CPTED aimed to promote the feeling of safety for the community, and vulnerability for the criminals. Combined with the *broken windows theory* (where the police crack down on minor crimes to inhibit more serious ones), active maintenance, graffiti removal, and street cleaning, it expressed ownership and control. For too long people had been brutalised by urban living, hurrying home, and locking their doors behind them. Here was an opportunity to reclaim the streets through natural guardianship, building a sense of community, and pushing back criminality. Its success depended more on architects than on police officers, and it tacitly revived the concept of collective security lost in a century of industrialisation. CPTED was credited with significant decreases in crime in housing estates, shopping centres, schools, and parking lots, and is now routinely woven into the fabric of modern urban design.

Although a 20th century term, a form of CPTED was used in Paris when the city was rebuilt in the 1850s. Prior to that, much of Paris was a densely populated, congested, disease-ridden, medieval slum. It was a haven for criminals and revolutionaries, easy to barricade and difficult for the police to control. It was hardly a destination for a romantic weekend.

The vision for Paris was to reflect the glory of France and create a sanitary and ordered environment. Mark Twain said of Baron Haussmann, the official in charge of the city's transformation, "He is annihilating the crooked streets and building, in their stead, noble boulevards as straight as an arrow—avenues which a cannonball could traverse from end to end without meeting an obstruction more irresistible than the flesh and bones of men. The mobs used to riot there, but they must seek another rallying point in future." The result could have been brutalist, but the genius of Paris is that it combines grace and elegance with natural surveillance.

Not quite Paris, but York: CPTED creates a feeling of natural surveillance.

Technical Surveillance

CCTV as a means of technical surveillance, as you would expect, followed the development of TV which was first broadcast in the 1930s. The earliest CCTV was used in WWII by the Germans to monitor the launch sites of V2 rockets, and by the US to monitor nuclear facilities. In the late 1940s CCTV became commercially available. But these early systems were expensive, the images looked like a snowstorm, and, as there was no viable recording technology, they required constant monitoring.

The potential of images to contribute to crime investigation, however, became clear in 1957 when a device known as *Photoguard*, was installed in the St Clair Savings Bank in Cleveland, Ohio. Soon after it was fitted, the bank was robbed by Steven Thomas and Wanda Di Cenzi, accompanied by their getaway driver Rose O'Donnell. All three were identified and convicted based on Photoguard images. It wasn't quite CCTV, but a still camera that took 14 images on a timer when activated by a teller using a secret switch. Here was proof that technology was effective, but the Photoguard was temperamental, and CCTV remained expensive.

The Met police experimented with CCTV to monitor crowds in London during the visit of the King of Thailand in 1960, but it took more than two decades before the technology was sufficiently cheap and reliable to become a common part of urban life. The introduction of magnetic tape with its slowly revolving reels meant that images could be recorded, reviewed, and used as evidence. But the tape was expensive, and it needed to be changed every couple of hours, then catalogued and secured in a climate-controlled cupboard.

In the 1970s after the arrival of colour CCTV, images were easier to interpret, and slightly less boring to watch. Around the same time, the video cassette recorder (VCR) was invented, which opened a market for home videos, and for the more efficient recording of CCTV images. VCR was easy to handle, and, to make the tapes last longer, they used time-lapse techniques. Rather than record continuously, an image was captured every few seconds, which greatly extended the life of the tape without diminishing image quality. The ability to record and play back clear CCTV images enabled a change in emphasis from passive monitoring to active investigation.

To help address rising crime, many British towns fitted large-scale local government operated CCTV systems. The first was Bournemouth in 1985 and within a decade they were common across the country. The Conservative government encouraged the spread of CCTV in its effort to promote itself as the party of law and order. In 1994 Home Secretary Michael Howard said he was "absolutely convinced that CCTV has a major part to play in helping detect and reduce crimes and to convict criminals."

Another major driver of CCTV rollout in the UK was IRA bomb attacks. The IRA found that an attack on civilians on mainland Britain generated much more media coverage than an attack on military targets in Northern Ireland. The used massive vehicle bombs in The City of London that caused widespread damage and threatened the viability of many businesses and The City itself. In response, a series of police checkpoints with CCTV surveillance, known as the *ring of steel*, was established to prevent vehicle bombs from being driven into The City.

The IRA also placed bombs on underground trains and in railway stations which led to the whole of London's transport network being placed under CCTV surveillance. Proud of its achievements, in 2002 Transport for London launched a poster campaign aimed at increasing public confidence in the security of its network. It featured the slogan *Secure Beneath Watchful Eyes* and an image of a red bus overlooked, not by a camera, but by an image of an all-seeing eye that might have been copied from a church pulpit. The *Watchful Eyes* continued to multiply and by 2020 there were more than 15,000 CCTV cameras in the underground network and each of London's 8,000 buses had 16 cameras.

Once the preserve of military installations, suddenly CCTV was everywhere: high streets, train stations, football grounds, shopping centres, offices, hotel receptions, roads, garage forecourts, buses, trains, concert halls, schools, airports, and ATMs. Camera technology improved. It became more compact, set in shiny fisheye domes, capable of panning, tilting, and zooming, and linked to infrared lights to see in the dark. With the arrival of digital technology, catatonic security officers no longer had to stare at screens. Systems could automatically detect motion, spot unattended luggage, identify loiterers, see crowds gathering, and focus in on scuffles. Their high-resolution images were no longer stored on spools of magnetic tape, they were stored on the cloud, instantly retrievable using digital analytics tools.

By 2020, across the UK there were more than 50,000 local authority CCTV cameras, each costing between £4,000 and £9,000 a year to maintain and operate. But the big expansion was in the commercial sector, with the British Security Industry Association estimating that by 2018 there were between 4 and 6 million privately operated CCTVs across the country.

CCTV has become the 5th utility after water, electricity, gas, and telecommunications. In 2000 it was said that the average Londoner was captured on camera 300 times a day. By 2020 it would be easier to count the moments when

Persuading the public that CCTV was a good thing.
(Courtesy of the London Transport Museum)

they were not actively stalked by CCTV. A tech review site, Comparitech.com, listed the world's most surveilled cities in 2020. Nine of the top ten were in China, but London was, depending on your point of view, a respectable or an alarming third, with one camera for every 14 people.

CCTV is not just peering down from a lamppost or a doorway, it is right in your face. In 2018 front-line Met Police officers were issued with body worn CCTV to record their interaction with the public. I asked a police officer how he felt about wearing a bodycam. "I love it," he said, complaints against me have dropped massively." He added, in a revealing moment of candour, "I must admit that I used to go in a bit hard, but I'm much more careful now." A 2021 study by the Royal College of Policing stated that complaints against the police had dropped by 16.6%. It's not clear if it was the police or the public who behaved better as a result of the close surveillance, but private security officers were soon wearing them too.

At the other extreme, military-grade airborne surveillance is now watching us from above. The UK police have been using drones for surveillance and to support their operations since 2014. Much of this was uncontroversial, but their use in 2020 by Derbyshire Police to enforce coronavirus restrictions in a national park, was widely criticised as excessively intrusive.

There were similar concerns in Baltimore in the US, when a private company, unnervingly named Persistent Surveillance Systems, was contracted to monitor activity in the city using an array of sensors on fixed-wing aircraft. The velvet-voiced narrator on its promotional video purrs, "It's not a spy in the sky, it's a tool to vindicate the innocent, identify the guilty and keep the police accountable." The first technical surveillance systems had their origins in the German V2 rocket programme of WWII; this latest one honed its capabilities in killer drones circling over Middle Eastern battlefields.

At least your own home remains a bastion of privacy. You can wear your pyjamas and drink beer all day without fear of being spotted by a camera. Or can you? Having saturated the market for CCTV in urban spaces, public transport and commercial premises, the final frontier was your home. The earliest home CCTV was patented in New York in 1969, by Marie and Albert Brown who were concerned about crime in their neighbourhood of Queens, New York. But it took another 20 years before the technology was sufficiently cheap, and easy to fit, for the domestic market to take off.

CCTV companies inevitably used the stoke and soothe technique to build sales. A simple question, "Is your home protected?" gnawed away at homeowners and their anxiety propelled them to the nearest DIY outlet where dozens of models are now on sale. By 2021 the global home CCTV market was $6.5 billion annually and growing at 15% a year. Having put a foot in the home market door, CCTV companies increased sales by offering high-definition cameras, recording options, door control systems, vaults for digital recorders, remote monitoring, service contracts, and security lights. You want your family to be as safe as possible right? Take no chances. If you really care for them, you buy all the extras.

Some companies offered doorbell videos that can be monitored remotely using a smartphone. The US firm *Ring*, for example, was a crowd-funded start-up in 2013, and rapidly expanded using the slogan "be in when you are out." The company was bought by Amazon in 2018 for $1.5 billion. In 2020, to encourage people to buy CCTVs for inside as well as outside their homes, it updated its slogan to "cover every corner, inside and out".

84 THE RISE OF SECURITY and Why We Always Want More

Be in when you're out. A Ring CCTV doorbell by Amazon.

The next iteration will surely be to link CCTV with Alexa, home delivery services, and facial recognition. Alexa will let you know as the named Amazon driver approaches. If you are out, you can see exactly where your parcel is left. And there will also be an option to upload the faces of your family and regular visitors, and for Alexa to issue alerts, "beware unknown person approaching." It could be a postman, a neighbour, a delivery from another company, or a thief.

Our attempts to monitor our homes will generate anxiety for an everyday situation to which we had previously paid little thought. Conveniently though, having stoked fear, Amazon will show you advertisements for locks and panic alarms and buying them will soothe your concerns. Is Amazon fulfilling your needs, or exploiting your paranoia?

Paranoia? A 2019 survey of 1,008 US homeowners by highsecurityhome.org, a US home security products review site, found that 40% of people were more anxious as a result of having a doorbell video. One respondent said, "I can't stop checking my phone every time the alert goes off. Person walking by? Check. Mail delivery? Check. Bird landing on our porch? Check. It's made me obsess over it. I feel like I always need to be watching now, not something I dealt with when it wasn't an option." That the products we buy to reduce anxiety end up fuelling it is the ultimate irony.

In the space of 80 years CCTV had progressed from providing grainy images of the most sensitive weapons, to pin-sharp images of deliveries to your door. CCTV proliferated for a variety of reasons: it was regarded as a useful crime prevention tool, it reassured the public, it was cheaper than a police officer, more alert than a security officer, and from being accepted in most places, it became expected everywhere.

CCTV is good for the security business, but is CCTV good for security? Although ubiquitous in shops, CCTV is a long way down the list of how to counter shoplifting. Active staff engagement with customers, well-lit shelves, posting shoplifting policies, locking high-value items in display cabinets, using electronic tagging, maintaining an up-to-date inventory (so stolen items can be identified), keeping a tidy shop, and restricting access to changing rooms, are all ahead of CCTV when it comes to preventing retail theft.

When it comes to urban crime, there is a case for CCTV. A 2009 review of 41 academic studies by Brandon Welsh and David Farringdon showed that the presence of CCTV reduced overall crime by 16%. It was especially effective in car parks where crime fell by 26%. But CCTV had negligible impact on violent crime. Theft is normally planned so the presence of CCTV diverts some criminals to areas without coverage. But violence tends to be emotional, reactive and spontaneous, so it is uninhibited by cameras. The major benefit of CCTV is after an event to help with investigations and to identify suspects, and it is used in 95% of all murder cases.

What about the impact of CCTV on terrorism? The IRA avoided The City after the *ring of steel* was built, choosing instead to bomb Canary Wharf and Manchester which did not, at the time, have the same level of surveillance. The IRA never went in for suicide attacks. They wished to fight another day, so they hid their identities and tended to be camera-shy. Islamist terrorists by contrast, aimed for martyrdom and all the good things that they thought came with it, were not inhibited by CCTV. On the contrary, knowing that CCTV images of their attacks will be broadcast to a shocked public they may, like reality TV stars, actually seek camera coverage.

No security professional would argue against installing CCTV. If there are no incidents it could be because CCTV acted as a deterrent. If there is an incident, CCTV may provide evidence. And if an incident is not caught on CCTV, it makes a case for the installation of more cameras. CCTV is no longer an optional extra, it is a fundamental and ubiquitous requirement.

But there are other perspectives beyond the realm of physical security. Emma Carr, Director of UK civil liberties campaign group Big Brother Watch claims, "Britain's crime rate is not significantly lower than comparable countries that do not have such vast surveillance." The concern is that individuals' rights to privacy are overlooked because of wider security benefits. Article 8 of the European Convention on Human Rights states that, *everyone has the right to respect for his private and family life, his home and his correspondence.* The Article balances this right against the *interests of national security, public safety or the economic well-being of the country*. It is hard to argue that personal liberty beats public safety. As Louis Freeh, Director of the FBI in the 1990s, said, "The American people must be willing to give up a degree of personal privacy in exchange for safety and security."

Your Face Becomes Your Fingerprint

Being seen is one thing, being identified is another. But identification is now possible using facial recognition technology that match faces seen on camera with those listed on digital databases. Your face becomes your fingerprint. Unlike fingerprints that need to touch a scanner, a face can be identified by a camera at a distance.

The technology has found increasing numbers of applications since it was developed by the US defence establishment in the 1990s and was first adopted by the US Department of Motor Vehicles to verify US driving licences. Established in 2004, Facebook softened the ground for acceptance of facial recognition technology. It was the first platform to make us shed our inhibitions about posting our images online. It may have started out as a US college jape to decide who was hot and was not, but it soon persuaded us to reveal our names and our faces to our friends and to Facebook's advertisers. By 2014 the company developed what it calls Deepface technology to scan photos and automatically identify any of the 2.8 billion people on its database.

In 2010 the UK issued biometric passports which allowed travellers to pass through automated border gates if their face matched their passport photo database. By 2017 the iPhone X could be opened with facial recognition, a further evolution from the pin code and finger scan. In Japan, facial recognition is used instead of tickets on some train lines. No problem if you forget your wallet, you always have your face with you, which can be matched to a database and used to debit your bank account, literally in the blink of an eye.

You might think the pandemic and the wearing of masks put a stick in the spokes of the facial recognition wheel. Wrong! The COVID-19 pandemic increased the demand for contactless interaction. It forced facial recognition to up its game and provide reliable identification of a masked face just by using data points around the eyes. Combined with thermal imaging it can even identify individuals who may have Coronavirus.

Much of the world is unaware, or unconcerned, that it is starring on real-life reality TV. But not San Francisco, home to many of the companies at the heart of surveillance technology. In 2019 the city banned the use of facial recognition by law enforcement agencies. What does it know that the rest of us don't?

Dataveillance

That China has enthusiastically embraced surveillance is partly a product of its geography. It is dominated by the Yellow and the Yangtze rivers which tumble from the Himalayas, and snake across vast alluvial plains, before oozing into the East China Sea. The rivers are a blessing and a curse. A blessing because agriculture in their fertile plains supports the highest density of people on earth. A curse because the rivers are incontinent, bursting their banks and changing course by up to 400 km in a single season, often causing massive devastation and loss of life.

To tame the rivers the Chinese needed monumental flood defences. Entire communities were mobilised to construct them, disciplined to maintain them, and committed to unquestioning obedience to operate them. This shaped the national character, making them value strong central authority and community cohesion over individual freedoms.

Fast forward hundreds of years and by 2019 China had deployed millions of facial recognition cameras that link a person's face with their car, home, workplace, on-line activity and friends. This data is integrated with a social credit system, where citizens are assigned a score based on their behaviour. If people pay their bills on time, they'll get a credit, if they drop litter, or drive too fast, they'll get a debit. A high credit rating may open opportunities for travel and for state jobs. A low score may result in public shaming, travel bans and limited access to bank loans. Given its geography, China was always lush territory for advanced surveillance and, after years of the *one-child policy*, no one could remember what a big brother was.

It sounds spooky but the UK isn't so far away from this. If you've ever glimpsed a white flash in your rear-view mirror as a camera catches you speeding, you know that a few days later a penalty notice will flop through your letterbox. Your name, car registration number and address are on a database. This could easily be linked to any number of other databases: passport, medical records, criminal records, social security, national insurance, fingerprints, electoral roll, DNA, sex offenders, tax, land registry, Companies House, credit ratings and GPS of your mobile phone, and monitored constantly. Upgrade government CCTV with facial recognition cameras, merge it with all these databases, and watch the falling rates of crime and illicit liaisons as all your activity is tracked, recorded and stored.

The US think tank The Carnegie Endowment for Peace, estimated in 2020 that at least seventy-five countries have advanced surveillance systems of varying levels of sophistication and penetration. Predictably these include authoritarian regimes, but many democratic ones are also getting with the programme. It's easy to suggest that authoritarians use surveillance for social control and that democracies use it for public security. But they are essentially the same thing: inhibiting deviation from lawful and moral conduct.

Security *is* ultimately about controlling people's behaviour. Traditionally this has been achieved through social, religious, and legal inhibitors. In recent times the first two have been diluted and the last one depends on law enforcement and justice which can be difficult, time-consuming, and expensive.

Advanced surveillance automates the monitoring of individuals and is cheaper and more reliable than humans. In the past, smart policing meant an officer with a nicely pressed uniform and shiny boots. Now it's a camera standing beside a traffic junction in all weathers without taking a lunch break, spotting cars jumping red lights.

The power and the pervasiveness of advanced surveillance both impresses and depresses. Some argue that you have nothing to fear, if you have nothing to hide. Others invoke the spirit of Orwell and complain about invasions of privacy, the undermining of trust, and the unreliability of technology and the people who operate them.

Whilst there is some discomfort about governments sucking up personal information, people are surprisingly willing to give details about every aspect of their lives to tech companies. David Omand, a former head of the UK government's eavesdropping service GCHQ, told an audience in 2018, "The big revelation over the last couple of years has been not about government intelligence agencies, it's been about the private sector. It is about the internet companies knowing more about me, you, everyone in the hall, than any intelligence agency ever could or should know about us."

Social media is dataveillance capitalism turning personal data into a commodity. Access to their platforms is free because, in exchange for being connected to your social network, your data is harvested and sold. As you click away, posting, commenting, and liking, Facebook gets to know you better than your mum and signals your preferences to retailers who then you target you with ads. So, a teenager gets ads for skateboards rather than wheelchairs, a pensioner gets ads for varicose vein treatment rather than acne medication, and an angler gets ads for fishing rods rather than football boots.

And yes, Facebook knows where you are. Even if you disable the location feature it can still use browsing habits and Wi-Fi connections to track your path.

We are increasingly willing to submit to surveillance in exchange for social connectivity. Older generations find this disturbing. Younger generations have fewer concerns. When dating, many young people will use a location-sharing app as an act of trust between couples. Trust, in a close relationship that you willingly enter is one thing, but can you trust your government, or your social media platform? And can you be confident that the data that they collect today will be secure from the hackers of tomorrow?

And what about your employer? Advanced surveillance is available to them to monitor your productivity. In February 2020 Barclays bank staff in Canary Wharf complained about the intrusive monitoring by a software called Sapience. It provides "automated work pattern reporting and real-time analytics" and "unprecedented visibility into how people work, with actionable insights to better manage cost and performance across teams." In essence, it created a modern-day labour camp panopticon without having to build a circular structure.

In an office you can see what people are doing, but with the trend for more home working, how can you tell if they are being productive or sharing funny cat videos on TikTok? The pandemic has been a boost to worker surveillance technology offered by companies with brand names such as Prodoscore, TimeDoctor, Hubstaff, and the one that tells it really straight: Staff Cop. These systems measure desk time, phone calls, email traffic, document data, and progress towards targets. They score each workers productivity to drive efficiency and therefore profit. The debate is whether it turns workers into data driven automatons, or provides an objective measure of their performance on which to base their rewards.

In 1954, when the all-seeing eye was little more than a motif on a church, American science fiction writer, Philip K Dick, penned a book called Minority Report which was turned into a movie by Steven Spielberg in 2002. The story was set in 2054 in an advanced technological society where surveillance had transitioned from observing, to identifying, to monitoring, to controlling human behaviour.

In Minority Report, the next step—anticipating human behaviour—is being made. The police have a PreCrime Department that predicts murders before they are committed, and it arrests people before they strike. We'll see in later chapters how security operations centres have moved from recording what has happened, to reporting what is happening in real time, and they now probe the frontier of predicting what will happen. The PreCrime Department no longer looks like science fiction.

As the tools of surveillance become increasingly sophisticated, our behaviour is under ever closer scrutiny. It feels as if we have become human salamanders held before a spotlight, our beating organs on display through translucent skin. There is no doubt that we are sacrificing our privacy and being forced to conform to parameters set by the surveillance operators, be they governments, transport systems, employers, or social media. But the survival of increasingly complex and populous societies relies on social control being maintained. Surveillance makes it harder to commit a crime and reduces the prospect of being the victim of a crime. The all-seeing eye forces us to trade individual liberty for collective security. It is scary, but it may not be all bad.

7

HIJACKS AND HIGH SECURITY

Think of flying and you might conjure an image of reclining in a comfortable seat sipping a gin and tonic, watching a movie, the world unfolding below as you speed to your destination. More likely you are wondering how long it will take to get through security, if your shampoo and nail clippers will be confiscated, if you'll have to take off your shoes and if so, whether there are holes in your socks. To an airline passenger, getting through security is a greater source of stress than having a stranger fall asleep on your shoulder and start to dribble.

It didn't used to be this way. Until the 1970s you could fly without pre-boarding checks, or any form of identification. You could smoke on a plane. You could visit the cockpit and admire the swanky pilot at work. Everyone watched the same in-flight movie, at the same time, from a screen that descended from the ceiling. The hostesses looked young and happy. Flying was so exciting and glamorous that Frank Sinatra eulogised it in his song, "Come fly with me". You used to feel like royalty. Now you feel like a high-risk offender entering a maximum-security prison.

In this chapter, we'll look at how aviation security evolved in response to threats from extortionists, asylum seekers, and terrorists. We'll consider the costs, the consequences, and the proportionality of the security measures, in what has become one of the safest ecosystems ever developed. Finally, we'll look at 9/11 and consider if security screening could have made a difference and how rational the response to the attacks was.

In 1970 there were 386 million air passenger journeys globally. By 2020 the numbers had risen to 4.7 billion. Flight had reached the masses. You couldn't smoke on board anymore, or visit the cockpit, but you had an individual flat-screen TV, and you could watch what you wanted, when you wanted. The hostesses had been joined by men, although none of them looked quite so young, or happy. And security procedures had become a dominant part of the flying experience.

Commercial aviation started after WWII with ex-military transport planes such as the Douglas DC3, their camouflage painted over with bright liveries, and their sideways canvas benches swapped for forward-facing foam seats. You exited via steps rather than by parachute.

In 1952 the British de Havilland Comet entered service as the first commercial jet airliner. Unfortunately, it had a habit of crashing and the jet age didn't really take off until the arrival of the Boeing 707 in 1958. This was also the first year that more passengers crossed the Atlantic by plane rather than by ship. In 1970 Boeing added the wide-bodied, 366-seater 747 to its fleet. It aimed to democratise air travel by reducing seat costs by a third. Airbus entered the market with the A300 in 1972. Boeing and Airbus continue to dominate the commercial aviation sector, having delivered 25,000 and 20,000 aircraft respectively over 50 years. De Haviland folded in 1963.

Let's have a quick look at aviation safety before returning to that queue at security. As we know, safety is about accidental or natural occurrences, whereas security is about intentional human acts, but the two are intrinsically linked. Between 1959 and 2019 more than 65,000 people worldwide were killed in commercial aviation safety incidents. That sounds like a lot until you compare it with the 60 million (yes million) or so killed on the roads during the same period.

In 1959, there were 40 fatalities per one million aircraft departures in the US. Within ten years this had dropped to fewer than two fatalities for every one million departures, falling to around 0.1 fatalities per million by 2020. So, flying became 400 times safer over 60 years. You are now many times more likely to be killed by lightning than in an aircraft. This is an extraordinary achievement when you consider that there are more than 100,000 flights worldwide, every day of the year.

Why has flying become so much safer? Inevitably, the more you do something the better you get at it. You learn from your mistakes. The reliability of aviation technology steadily improved, inspection and compliance regimes became more thorough, air traffic was more tightly controlled, the human risk factors were more closely managed, and there was better crew training. Improving aviation safety was fairly straightforward. In contrast, aviation security had to account for human cunning and ingenuity, which is much harder

A Brief History of Hijacks

Aviation opened up an entirely new security arena. Hijacking ships—piracy—has been around since the advent of sea faring. While pirates were motivated by money, the motivation for aircraft hijackers went through a series of phases: firstly, it was about forcing an aircraft to divert to a particular destination, then it became about money, then, once the publicity value became understood, it became about politics—in other words, terrorism.

The first hijackers just wanted to get somewhere. In the 1950s and 60s planes were hijacked by people escaping communist regimes in Romania, Yugoslavia, Hungary and Czechoslovakia. The Cuban revolution spawned two-way hijacking traffic; people escaping the Cuban regime, and Cubans in the US wishing to return home after transport links between the two countries were severed. Between 1968 and 1972, 90 aircraft were hijacked in the US and flown to Cuba. It was so common and untragic that Monty Python parodied hijacking in a sketch called "Take me to Cuba".

By the 1970s there was a spate of hijackings for money. The strangest case was that of DB Johnson, the pseudonym of a man who in 1971 hijacked a flight from Portland, Oregon, to Seattle. Claiming to have a bomb in his briefcase, he negotiated the release of the passengers in Seattle in exchange for $200,000 and a parachute. He then instructed the plane to fly to Reno, with just the crew and himself aboard. After strapping on the parachute, "DB Johnson" jumped out somewhere over Washington State. He was never seen again, but he inspired others: in the following year, there were 12 copycat hijacking attempts by people using parachutes. All were unsuccessful.

Airlines initially had a policy of total compliance with hijackers' demands. Many even kept charts of Cuba in the cockpit in case they were diverted. As hijacks seldom resulted in violence or damage, the philosophy was to avoid publicity, which might dent enthusiasm for air travel, and to accept the inconvenience of diverted flights. Hijackers were not motivated to kill people and they did not want to get killed themselves, so police interventions were normally negotiations rather than shoot-outs. Also, air travel was a new, sexy business, with deep roots in the US Air Force (freshly triumphant from a World War) and in space exploration (the Apollo programme mesmerised the world from the first mission in 1961 through to the moon landing in 1969). So, aviation had plenty of political clout and it rode a wave of goodwill from a public who saw it as a heroic, pioneering endeavour. A few hijacks were not going to hold back the adventure.

Brendan Koerner, author of a 2013 book on hijacking, The Skies Belong to Us, points out that in the 1960s people were more phlegmatic and more likely to think whilst being hijacked, "I'll spend the night in Havana, I'll have a story to tell at the next cocktail party, and maybe I'll smuggle back some cigars and rum." Today, they would call a lawyer for compensation, a counsellor for the stress caused, and an agent for a book deal.

But things changed in the 1970s. Along with mass air travel, mass television ownership took off. In the UK by 1970, two years after the introduction of colour, 90% of households had TVs. Hijacking, meanwhile, had all the essential ingredients for a big, engrossing news story: danger on the newest, most glamorous form of travel, an international setting, agonising decisions, men with guns, innocent victims, perhaps a spectacular explosion, and the prospect of multiple casualties. The newsworthy nature of hijacks became the principle motivation for terrorists who sought publicity to change people's attitudes and government policy.

While real life made news on the small screen, Hollywood soon recognised the opportunities that hijacks presented for gripping entertainment on the big screen.

The 1970s was a golden era for disaster movies and everyone of a certain age will recall *The Poseidon Adventure*, *The Towering Inferno* and *Earthquake*. The Daddy of them all, released in 1970, was *Airport*, about a plot to blow up an airliner in flight. It was so successful that it spawned several sequels: *Airport 75* (small plane flies into the cockpit of a 747); *Airport 77* (hijacked plane disappears in the Bermuda Triangle); *Airport 79* (surface-to-air missile plot against an American Concorde); and even the parody *Airplane!* (crew incapacitated by food poisoning leaving passengers to land the plane). The genre swelled with classics such as: *Air Force One*, *Con Air*, *Executive Decision*, *Skyjacked*, *Passenger 57*, and *The Delta Force*. Like all disaster movies they featured tension, spectacular effects, and square-jawed heroes including Burt Lancaster, Chuck Norris, Charlton Heston, Wesley Snipes, and Harrison Ford.

These movies were popular because they were intrinsically entertaining, but they were also relatable, credible, and instructive. Relatable, because by 1970 many people in the West had flown. Credible, because despite airlines' attempts to play them down, hijacks became news staples. Instructive, because we are naturally fascinated by how people react under stress, and we wonder how we would cope in similar circumstances ourselves. Hijack movies pose questions for every viewer: would you stay calm under pressure, wrestle with the attackers, help the injured, or be a whimpering mess?

The first major political hijacking, the longest—and the only time an El Al aircraft has been taken over—was in 1968 when a flight from Rome to Tel Aviv was seized by the PFLP (Popular Front for the Liberation of Palestine). The plane was flown to Algiers where it stayed on the runway for 40 days while diplomats negotiated the release of Palestinian prisoners in exchange for passengers. It was never made into a movie, but it dominated TV news broadcasts worldwide for more than a month, which was exactly the publicity the PFLP sought.

Emboldened by the El Al hijack, in 1970 the PFLP seized four aircraft simultaneously (one each from BOAC, TWA, Swissair, and Pan Am) and flew them to Dawson's Field, a former Royal Air Force base in Jordan. The incident lasted two weeks, no hostages were killed, but the planes were blown up by the hijackers, and seven Palestinian prisoners were released.

The overseas hijacking of US airlines spurred President Nixon into action. A few low-profile hijackings to Cuba were one thing, but it was quite a different matter when US citizens, on US aircraft, were being held for ransom in foreign lands, with the saga beamed into people's living rooms night after night. The US had just put a man on the moon, it wasn't going to be humiliated by a handful of terrorists on a desert airstrip.

Nixon met with many of the 191 freed US passengers and crew after their release from Dawson's Field. He told reporters, "the possibility of reducing hijackings in the future had been substantially increased, because the international community was outraged by these incidents. Now we have not only mobilized guards on our planes, but we are developing facilities ... for the purpose of seeing that people who might be potential hijackers do not get on planes with weapons or explosive material."

It was not just the US that was concerned about hijacking. It had become an international issue of critical proportions. During the Dawson's Field incident, the UN Security Council adopted Resolution 286 calling upon States to take all possible legal steps to prevent further hijackings. In December 1970, 77 States gathered in The Hague to sign the Convention for the Suppression of Unlawful Seizure of Aircraft. This provided a legal framework to deter hijacking and to encourage international cooperation. It was followed by the 1971 Montreal Convention for the Suppression of Unlawful Acts against the Safety of

Civil Aviation which criminalised acts of violence on-board aircraft, along with destroying or damaging them.

To implement these new policies, practical steps were needed. The International Civil Aviation Organisation (ICAO), a specialist UN agency established in 1947 to coordinate international air travel, led the way. The 1975 Annex 17 to the Convention on International Civil Aviation required States to "establish and implement a written national civil aviation security programme to safe-guard civil aviation operations". It laid out standards for access to aircraft, the screening of cabin and hold baggage, and access to the flight crew compartment.

While US airlines had initially taken a discreet, acquiescent approach to aviation security, in the late 1960s the Federal Aviation Authority took its first steps to control the rash of hijacks. It developed a passenger profiling programme to identify people who might seek to go to Cuba rather than the plane's scheduled destination, and it established a small cadre of Sky Marshals. These were discretely armed guards aboard planes at higher risk of hijack. The guards were never called upon to respond to an incident and after a few years the programme slipped into obscurity until, as we'll see, 9/11.

From the 1970s, security became a prominent feature of air travel worldwide. The most visible aspect was X-Ray screening for cabin baggage, accompanied by a desultory pat down, or a wave of a metal detector wand. It was a casual rather than a comprehensive regime, but it had an effect as the number of hijacks halved during the decade.

With hijacking aircraft becoming more difficult in the 1970s, terrorists switched to attacking airports including: Tel Aviv, New York, Paris, and Rome. Passengers were searched before boarding aircraft, but airport buildings were open to all, without any access control. This shift in targeting underlines how security threats are adaptive; if you close down one vulnerability, bad guys seek alternatives.

Airports were easier to attack than aircraft, but they didn't have the same impact as airborne drama. People in airports weren't as easy to take hostage, there were armed police who could respond rapidly, so incidents were soon resolved, and the body count was lower.

Terrorists shifted tactics again. Being made of metal, guns are easily picked up by X-Rays and metal detectors, but bombs are much easier to get past security. They can be hidden in anything from a book to the lining of a suitcase. Their component parts are an explosive charge, a detonator to make the explosives go bang, and an initiation mechanism.

Chapter 7: Hijacks and High Security **97**

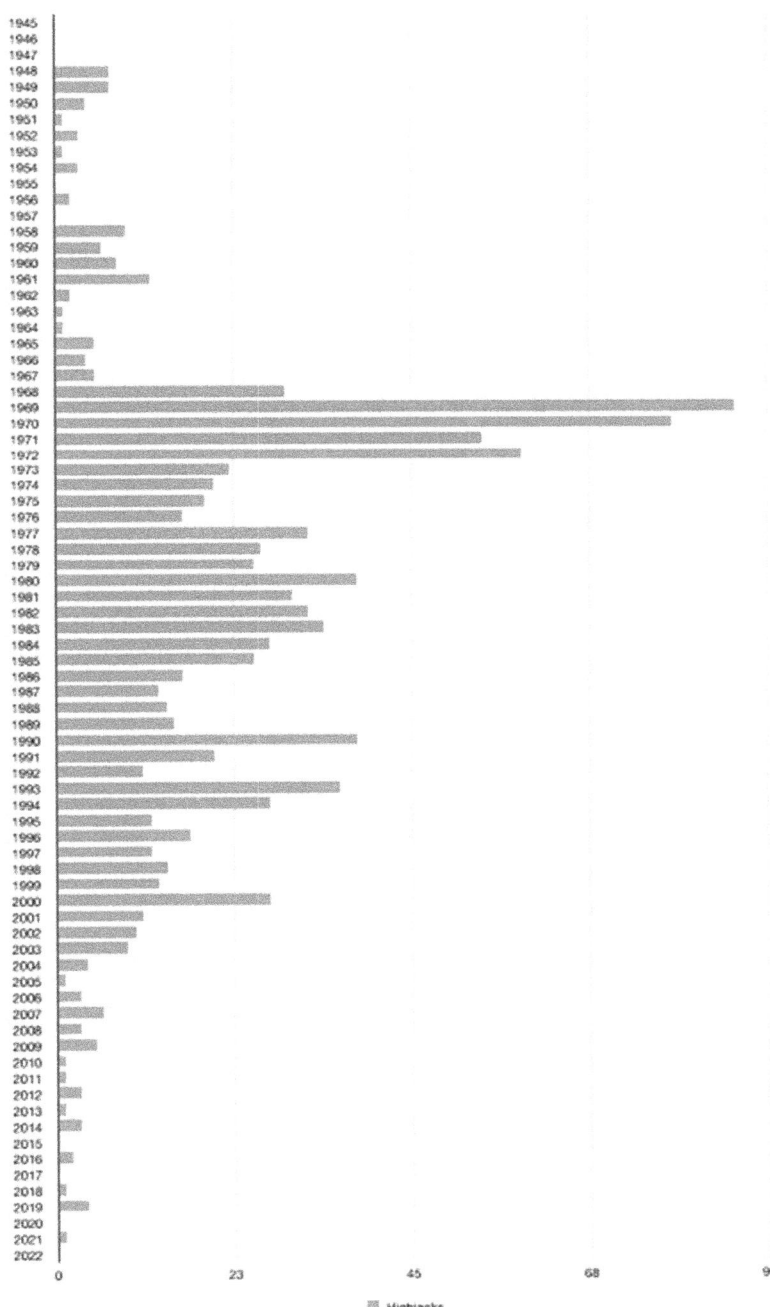

The number of hijacks globally since 1945.
(Source Aviation Safety Network Database)

The amount of explosive needed to rupture an aircraft's hull with catastrophic consequences is alarmingly small, especially at altitude where there is a big difference between the cabin pressure and the outside pressure. The detonator is an alloy tube, smaller than a pen cap, containing a gram of explosive compound. There are two varieties, a flash detonator that is initiated by a spark from a fuze lit by a match, and an electrical detonator with protruding wires which is initiated by current from a battery. The initiating mechanism, to close the electrical circuit between the battery and the detonator can be a simple button or switch, a timer, a barometric pressure sensor, a light sensor, or a mechanism that responds to changes in angle.

A bomb can be carried onto an aircraft intact, or in its constituent parts and then assembled on board. As bombs are easily concealed and come in random shapes and sizes, it is hard to train security staff to spot them. So, getting a bomb on an aircraft was no great challenge.

A terrorist with a bomb can either detonate it and kill himself and everyone on board, or he can threaten to use it to demand a political concession. In the 1960s and 70s there were a handful of aircraft losses that may have been caused by a bomb. But investigations were inconclusive, either because the forensic techniques of the era were limited, or the evidence was scattered over oceans or deserts and was hard to retrieve.

In 1976 a bomb detonated aboard a Cedar Airlines flight from Beirut to Dubai, killing all 81 passengers and crew over Saudi Arabia. Seven years later, a bomb on board a Gulf Air flight from Karachi to Abu Dhabi was blown up mid-air killing 112. Being Middle Eastern airlines, with Middle Eastern passengers, downed in the Middle East, with wreckage lost in the desert or at sea, the reaction to these events was muted.

Then, in 1985 an Air India 747 en-route from Montreal to London was blown up mid-Atlantic killing all 329, mostly Canadians, on board. It was, until 9/11, the largest loss of life in an aviation terrorist incident. Three years later, in the skies over Lockerbie, Scotland, a Pan Am 747 was downed by a bomb and killed all 259, mostly Americans, on board, and a further 11 on the ground. The shocking image of the iconic 747's nose in Pan Am colours shattered on the ground was shared around the world and provoked demands for action.

Stopping the Terrorists

But what to do? Terrorists were now using hard-to-detect plastic explosives, and commonly available electronic components. Hitherto, bomb timers had been mechanical, using the hands of a clock or a watch to complete the circuit between a battery and a detonator, which meant that they could only be set up to 12 hours in advance. Like all things mechanical, the reliability of clocks was affected by variations in temperature, humidity, altitude, and angle. And their ticking was also a give-away.

By the 1980s electronic timers, like those mass-produced for home video recorders, were readily available. They were much smaller and more dependable than mechanical timepieces, they could be set to detonate days in advance, and they functioned silently. Another option, as used in the Lockerbie bomb, was a small electronic barometric pressure sensor of the type used in car engine management systems. It could be set to initiate at a given altitude.

In the same way that the best place to hide a pebble is on a beach, the best place to hide a bomb with electronic components is within an electronic device. The Lockerbie bomb was concealed in a Toshiba cassette player and contained only 400 grams (about the size of two cigarette packets) of the plastic explosive Semtex. It was placed inside checked baggage, but the ticketed passenger did not board the aircraft. On detonation, at 30,000 feet, the bomb punched a 50cm hole in the plane's hull causing its immediate disintegration, and sprawling the wreckage over 700 square miles of Scotland.

To prevent this type of bombing, a combination of techniques was needed: increased intelligence efforts, ensuring that passengers boarded the plane with their checked baggage, improved screening of checked baggage, and making plastic explosives more detectable.

Screening checked baggage gradually became more sophisticated as X-Ray and explosive-tracing technology improved. Yet these were no silver bullets. The components of a bomb can be easily disguised, the volume of baggage on a flight is enormous, and the time available for screening is short. Keeping baggage bomb-free on 100,000 flights a day across the world was a huge challenge, as was discovered in 1990 when Jim Swire, the father of one of the Lockerbie victims, built a replica of the Toshiba cassette player bomb and succeeded in getting it through security checks on a London-New York flight. You can imagine the controversy.

You may well have been on a plane yourself, waiting to take off when the pilot's frustrated voice says, "I'm sorry to inform you that our flight will be delayed. A passenger has checked-in luggage but has not boarded the plane. Thank you for your patience whilst we unload their bags and we'll be airborne as soon as we can." Groans all round, but that procedure has been in place since Lockerbie.

The detectability of plastic explosives was addressed by ICAO, which met in 1989 to draft an international agreement known as the Convention for the Marking of Plastic Explosives for the Purpose of Detection. Under its terms, states agreed to tightly control access to explosives, and to use chemical tags to make them easier to identify.

So, how have terrorists tried to overcome increasingly rigorous airport screening? One of the main tactics has been to use suicide bombers with well-disguised devices and simple initiation mechanisms. In December 2001 a British Islamist fundamentalist, Richard Reid, boarded an American Airlines plane in Paris bound for Miami with 200 grams of plastic explosive concealed in his shoe. Mid-Atlantic, an air stewardess saw him lighting a match and told him that smoking was not permitted on board. Air France had only banned smoking in the cabin the previous year, so Reid's actions were not exceptional for the time. A few minutes later he again lit a match and tried to ignite a fuse to detonate the explosives, but, dampened by sweat from his feet, it would not burn. Fellow passengers then became suspicious and overpowered the would-be bomber.

The device was simple: plastic explosive and a flash detonator. Boarding an aircraft with this sort of bomb was, and remains, very simple. The security response was inevitable: passengers had to remove their shoes for an X-Ray inspection. The queues at security grew ever longer, and passengers' patience was further tested.

An Intelligence-Led Response

The Lockerbie bombing was a catalyst for increased US intelligence efforts. Further investment followed the truck bombing of the World Trade Centre in 1993, and the simultaneous bombings of US Embassies in Kenya and Tanzania in 1998 by Al Qaeda. In the following decade the 2005 7/7 London bombings had a similar impact on British intelligence efforts.

This paid off in 2006 when a group of British-based Pakistani men were arrested whilst plotting to simultaneously destroy seven US commercial aircraft en-route from London. Aware that screening might detect conventional explosives, the group planned to board the aircraft with bombs made from liquid hydrogen peroxide disguised in soft drink bottles. Once over the Atlantic, they would insert a small electric detonator and blow up themselves and all on board.

The plot was discovered and an immediate change in airport security procedures was announced: passengers were forbidden from carrying any liquids or gels (although this was later relaxed for containers of 100ml or less). Hundreds of flights were cancelled while airports attempted to put in place the new regulations, queues for enhanced screening snaked out into parking lots, and confiscated bottles generated mountains of plastic waste.

"Come Fly With Me." The joy of air travel.

The cost of the additional security measures was immense. Immediately, more than 400,000 passengers had their journeys cancelled; longer term the additional measures demanded yet more security staff. Let's say that 1,000 international airports (of the 15,000 airports globally that can receive international flights), each employing hundreds of security officers, increased their numbers by 20%, then the total uplift would have been in the tens of thousands of officers costing perhaps an additional half a billion dollars a year, every year.

The more stringent screening mechanisms further reduced the number of incidents, but terrorists were still able to adapt their tactics. On Christmas Day 2009, Nigerian-born Umar Farouk Abdulmutallab boarded a Northwest Airlines flight from Amsterdam bound for Detroit. On the approach to Detroit, Abdulmutallab attempted to ignite a device similar to that used by Richard

Reid, concealed not in his shoe, but in his underwear. The device caught fire, Abdulmutallab was overpowered by passengers, the flames were extinguished, and the plane landed safely. He was later sentenced to life imprisonment and he earned himself the title of The Underpants Bomber.

This attack hastened the deployment of body scanning technology that had been around since the early 1990s. Body scanners detect objects on, or within, a person's body without them having to remove their clothes and, unlike metal detectors, they can identify non-metallic items. These scanners use either backscatter X-Rays, or millimetre wave scanners. Backscatter X-Rays detect variations in hard and soft materials using radiation which reflects from an object and forms an image. Millimetre wave scanners create a similar image using high-frequency radio waves. In contrast, traditional X-Rays look through an object to create an image and are harmful to human tissue.

The adoption rate of body scanning technology was slow due to concerns about health and safety and the prospect of operators viewing images of passengers' naked bodies. Some felt it was like having an electronic strip search. And they had a point. Early versions of the full body scanners were revealing. In May 2010, Rolando Negrin, a security officer at Miami International Airport, was charged with assaulting a fellow security officer who mocked his genitalia that were displayed on a scanner during a training exercise.

Despite this, the Underpants Bomber incident caused a step change in the rollout of body scanners, although they were adapted to preserve passengers' dignity. In the following decade, they became a common feature in many airports, helping to speed up search procedures and improve detection capabilities.

It was getting increasingly difficult to get a bomb aboard an aircraft, so terrorists again shifted tactics and attempted to use missiles.

During WWII, guns had been used to engage aircraft. The chances of getting a direct hit at altitude were very low so anti-aircraft ammunition was designed to detonate at a set height, into a cloud of metal shards—or flak—to pierce an aircraft's skin. Even so, it was an inexact science, aimed principally at forcing bombers to fly higher to make their aim less accurate. German air defence units fired around 3,000 rounds for every allied bomber shot down.

Guns had no chance of hitting the new high-speed, high-altitude, military jets that were deployed in the 1950s. To hit a jet, you needed an even faster jet-propelled missile, with a guidance mechanism that could identify the heat from engines and zero in on them. The first real test was during the Vietnam war where hundreds of US combat jet aircraft were shot down by Vietnamese-operated and Soviet-supplied surface-to-air missiles (SAMs).

It was a 1970s Russian SAM known as a Buk that shot down Malaysian Airline MH17 as it flew at 33,000 feet over eastern Ukraine in 2014.

To reduce vulnerability to high-altitude SAMs, military aircraft tactics changed to low-flying contour-hugging missions. Inevitably, weapons manufacturers responded by developing short-range missiles to take out aircraft flying under 3,500 metres. These were light enough to be carried and operated by a single soldier and they became known as MANPADS: man-portable air defence systems.

Since the 1970s, about 30 civilian aircraft have been shot down using MANPADS, almost exclusively in conflict zones, most of them in Africa. In 2002 one was used in an attempt to shoot down an Israeli tourist charter aircraft soon after it took off from Mombasa. It caused only superficial damage, and no one was harmed.

The supply of MANPADS was strictly controlled, but many originating from conflict zones—Afghanistan, Bosnia, Iraq, Libya, Somalia—found their way into terrorists' hands. To guard against them, the area around an airport must be firmly controlled and intelligence networks have to be effective. These might appear relatively modest obstacles to the determined terrorist. So why haven't they been used more often?

There are several factors. MANPAD operators need specialist training, the custom-made battery packs run down and are not easily recharged or replaced, and their sale and deployment is tightly controlled. Following the Mombasa attack, leaders at the 2003 G-8 Evian Summit agreed to a US-initiated MANPADS Action Plan that included: programmes to destroy excess MANPADS and stringent export controls on MANPADS and their components.

The operational constraints and the non-proliferation initiatives were significant inhibitors. Another is that taking out a civilian airliner inevitably kills innocents which can undermine terrorists' objectives and reputation even among their own supporters. Furthermore, attacking an aircraft provokes an extreme reaction, as the 9/11 attackers discovered, and that is sufficient to deter many. A terrorist has bargaining power and kudos if he has a MANPAD, but he loses both as soon as he fires a missile, and he has the hounds of hell after him for the rest of his life.

And Then There was 9/11…

"We have some planes". These ominous words announced the start of the attacks that became known as 9/11. The details are well known: four planes were hijacked by Al Qaeda terrorists on September 11th 2001. Two were flown into the twin towers of New York's World Trade Centre, one was flown into the Pentagon, and the fourth was apparently planning to hit the White House but crashed in Pennsylvania after a heroic intervention by passengers.

The 9/11 plot involved 19 hijackers, aged between 20 and 33, 15 of them from Saudi Arabia, two from UAE, and one each from Lebanon and Egypt. They all had US visas and had passed undetected through US airport screening. They weren't carrying guns or explosives, and there was no evidence that any of the hijackers had weapons that were prohibited on board airlines. They used their muscle, aggression, and cunning, to stage the world's biggest terrorist attack.

We'll never know exactly how they took control of the cockpits. The only recorded evidence was of someone in the cockpit of United 93, the plane that crashed in Pennsylvania, shouting, "Hey, get out of here." The hijackers, who had seats close to the front of the plane, may have waited for the cockpit door to be opened, perhaps by a pilot going to the toilet, or by cabin crew delivering coffee. Perhaps they used improvised weapons to threaten the crew or the passengers. Either way, they took control of the aircraft remarkably swiftly and efficiently. The trained pilots amongst them then used the aircraft as guided missiles, killing 2,977 people.

The key single factor in the attacks was cockpit access control. The hijackers gained entry before any warning could be given, so they were certainly quick. At the time, the cockpit doors had to be locked on take-off and landing, but they were opened at the crew's discretion at other times. El Al planes had long had reinforced doors with strict protocols for opening them, but few airlines had similar standards and most cockpit doors, while lockable, were designed to be breakable so they could be accessed in an emergency. Following 9/11, the El Al approach was widely adopted.

There was no evidence that airport security was breached by the 9/11 hijackers. It is possible that they boarded with box-cutter knives (slim, razor-sharp, retractable blades within a hard-plastic sheath) but at the time you could carry blades up to 4 inches long on an aircraft. They may have used sharpened pencils pressed into the ear, or the throat, of a flight attendant to coerce them into opening the cockpit door. Whatever happened, the attacks were a catalyst for major changes in the administration of US airline security.

Until 9/11 airport security was contracted out to private companies that would supply staff for passenger screening. Their jobs, like so many in security, were repetitive and boring, often with unsocial shifts. Being low-skilled work, the staff were on little more than minimum wage. Staff might change jobs for an extra 50c an hour, so turnover rates were very high. Each security officer was screening as many as 1,000 people a day, a dehumanising experience for them and for passengers who were herded like cattle through the process.

Within weeks of 9/11, President George W. Bush signed the Aviation and Transport Security Act 2001 which established the Transport Security Administration (TSA), a single federal agency that assumed control of all aspects of aviation security. Initially it was under the Department of Transportation before moving to the newly formed Department of Homeland Security in 2003. By 2020 its budget had grown to $7.8 billion.

The TSA took a systems approach with 20 layers of security including passenger screening, behaviour detection, baggage checks and crew vetting. None of this was new but all layers, in the wake of the appalling tragedy, were upgraded and coordinated.

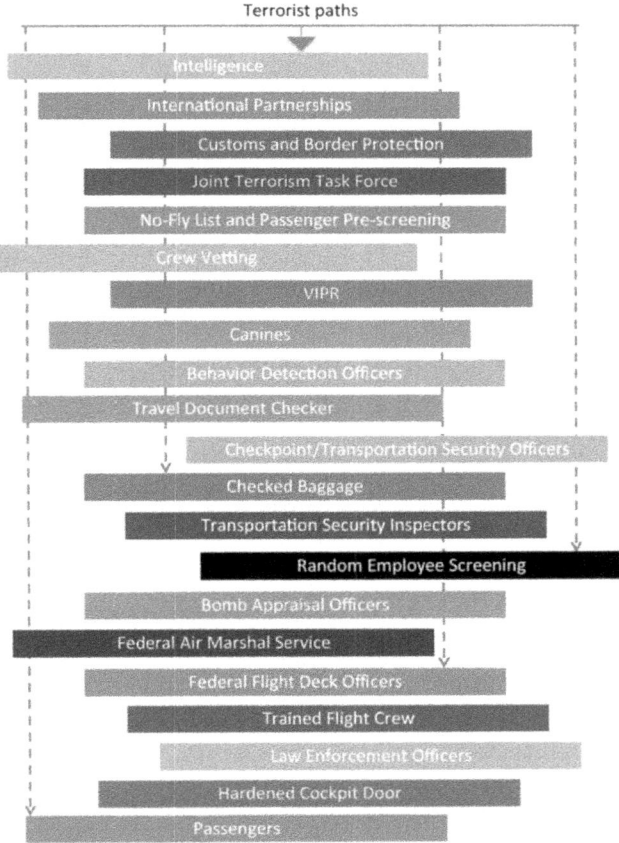

Passengers fighting back is one of the TSA's 20 layers of security.
(Courtesy of the TSA)

From the 20,000 or so airport security staff in the US before 9/11, following the establishment of the TSA, their numbers swelled to 45,000. Between February and December 2002, 1.7 million applicants were assessed for the new TSA jobs. This was the most intensive large-scale staffing project in the US since the mass military recruitment of WWII.

Although their numbers soared, entry benchmarks for airport security staff remained low. Applicants had to be US Citizens, at least 18 years of age, pass a drug screening test, medical evaluation, and background investigation, and either have a high school diploma or a year of work experience in the security industry. Unlike the private security companies who paid just above minimum wage, TSA Transport Security Officers (TSOs) were well paid by the sector's standards. In Kansas in 2020, for example, the minimum wage was $7.25 per hour but a TSOs' starting pay was $18.79 plus a 15% retention allowance.

Higher pay, combined with a smart uniform, comprehensive training, and a boost to their status, improved performance and morale. But few grow up with an ambition to be a TSO: the work remains dull and repetitive, and every day they process a thousand sullen passengers, attempting to maintain their dignity as they clutch their beltless trousers. It does not always feel like a noble and valued occupation, and despite the retention allowance a fifth of new hires quit in the first 6 months.

The Aviation and Transport Security Act also restricted access to departure areas to ticket passengers only, ending tearful farewells and cheerful hellos at the gate. It also instituted upgrades to passenger profiling using advanced computing to provide information on those considered a potential security risk and generated no-fly lists.

The other big upgrade was to the Sky Marshall Service. You'll recall that Marshalls had been posted on a handful of US aircraft since the Cuban hijackings of the 1960s. They received a modest boost from Richard Nixon following the Dawson's Field incident, but their numbers dwindled and by 2001 only 33 remained operational. Under the TSA, their numbers were rapidly boosted to around 4,000 which, by 2020, cost $1 billion a year. In addition to the direct costs, Sky Marshalls normally fly first-class to be close to the cockpit. Seats provided gratis by the airlines, which represents $200 million a year in lost revenue.

How Much to Feel Secure?

Sky Marshalls have not been without controversy. They have yet to intervene in a terrorist-related incident. The only person ever shot dead by a Sky

Marshall was unarmed. Having left an aircraft in Miami after an argument with his wife, he was killed in the airport terminal.

It is rare that politicians call for reductions in security. It is not a vote-winner, especially in the US after 9/11. However, Tennessee Congressman John J Duncan was vocal in his criticism, calculating that the 4,000 Sky Marshalls intervene, on average, in 4 non-terrorist incidents related a year, which works out at around $200 million per arrest.

A cost-benefit analysis into aviation security by academics Mark Stewart and John Mueller found that pre-boarding security reduced risks by 50%, crew and passenger resistance by 16.67%, hardened cockpit doors by 16.67%, and Sky Marshalls by 1.67%. Hardening cockpit doors cost $40m; Sky Marshalls cost $1 billion.

You'll note that Stewart and Mueller consider passenger intervention as effective at reducing the risk of hijack as reinforced cockpit doors. The 9/11 attackers ruined the market for ordinary hijackers. No one on an aircraft will ever again believe the words "Just be quiet and everything will be ok." On United 93, when passengers discovered that the first two hijacked planes had been flown into the World Trade Centre, they set upon the hijackers with such ferocity that the hijackers deliberately nose-dived the plane into the ground more than one hundred miles from its intended target. Quiet compliance has been replaced by aggressive confrontation.

Since 9/11 there have been no hijackings in the US and only 17 globally; in these incidents no passengers were killed, and no planes were damaged. Between 1987 and 2001 there had been only one hijacking in the US: in 1994, in what could have been a horror movie plot, a disgruntled employee attacked the crew of a FedEx cargo plane with a hammer and a speargun after it left Memphis bound for San Jose. The three crew members were severely injured, but they eventually subdued the attacker and landed the aircraft. So, US aviation security was a ready working before 9/11. But with security measures, you never know if you have deterred an attack, so it's impossible to precisely calibrate measures against threats.

Since 9/11 there have been no aviation bombing incidents in the US or on US aircraft, although elsewhere in the world there have been 8 attacks that killed a total of 418, an average of 20 a year. The largest single loss of life was on a Russian charter aircraft that was blown-up over Sinai after leaving Sharm el-Sheikh in Egypt bound for St Petersburg, killing 217 passengers.

On the safety side, globally since 9/11, 10,815 people have been killed in commercial aviation accidents, an average of 583 a year. This means that you are almost 30 times more likely to die in a safety incident than a security incident.

In the US, the figures are even more striking. In the past two decades there have been more than 14 billion scheduled air passenger journeys which resulted in 158 safety related deaths on board aircraft. In the same period, more than 720,000 were killed on US roads. So flying is many times safer than driving.

How much has it cost to achieve this level of security? It's hard to be precise as figures for the larger airport buildings needed to accommodate more extensive screening areas, and the additional time taken to navigate airport security (before 9/11 the average security checkpoint processed 350 passengers per hour, in 2020 the figure was under 150 passengers per hour), are hard to calculate. But in the US, a levy known as the September 11 Security Fee, is collected from each passenger as part of their air fare and remitted to the TSA. In 2020 the fee was $5.60 per flight and it covered $4 billion of the TSA's $7.8 billion annual budget. If these figures are extrapolated for air passengers globally the annual cost of aviation security is in excess of $20 bn, plus the disposal and replacement costs of all those confiscated shampoo bottles and nail clippers.

The decline in security incidents as more measures were introduced points towards their effectiveness. Or does it? During a series of penetration tests (attempts by covert TSA officials to breach security) in 2005, 67 of 70 Homeland Security Red Team members were able to get weapons though checkpoints: a 95% failure rate. In 2019, TSA officers seized 4,432 guns at checkpoints, which makes you wonder if this represents 5% of the total that people attempted to get through security. TSA Administrator David Pekoske said, "The continued increase in the number of firearms that travellers bring to airport checkpoints is deeply troubling." In the US it isn't illegal to bring firearms to the airport, but people must declare them, keep them in a locked case, and have them checked in. Reassuringly, the owners of the seized weapons apparently did not have hostile intent but were exercising their 2nd amendment rights, unaware of the procedures for flying with weapons.

It's impossible to prove that the additional security has prevented any attacks since 9/11. Security agencies are notoriously tight-lipped about their activities; however, the criminal justice system is less inhibited. Apart from the 2006 liquid bomber plot intercepted by British Police, there have been no trials of people intending to hijack or bomb an aircraft. The TSA caught an Army Vet in 2008 with bomb-making equipment in his suitcase that he was planning to transport to Jamaica, rather than using to attack the plane. As they publicised that success, we can assume that they would publicise others—had they occurred. But there has been a steady decline in incidents since the 1960s as more thorough screening procedures were put in place. So, we can assume that the cumulative efforts over the years have been effective. And a lack of arrests, whilst not a performance indicator, is a sign that security has a deterrent effect.

It is possible that there is some overspending on aviation security, but you cannot calibrate spending to ensure that precisely nothing happens. Security is not an exact science. It's hard to judge if you are paying more than you should. And even so, it is often better to overspend than to underperform.

There may be a case for reducing security screening for children and the elderly, and there are several "trusted traveller" programmes that attempt to make the boarding experience less onerous for some categories of regular travellers. These endeavours sometimes blend into issues around passenger profiling, which involves identifying higher-risk passengers and those behaving suspiciously and subjecting them to more intense scrutiny. This has inevitably led to concerns over racial stereotyping, discrimination and stigmatisation. Civil liberties groups have been vocal in their opposition, claiming that "flying whilst Arab" has become the new "driving whilst black".

Finding the right balance between security, resources, and liberty is no easy task. But sometimes it helps to consider how far we have come. Until 1958, most transatlantic journeys were by ship which took about a week. By plane it now takes 8 hours, including the security screening. As a result of safety and security innovation, flying is now, by far, the safest form of transport. Spending $20 billion annually on airline security seems an extraordinary price to pay to make death on an aircraft many times less likely than death by lightning. But it works out at $5 per person per flight. Most would happily pay this, and take their shoes off, for the additional peace of mind that it provides. And the saying in the business is: there is always too much security, until there isn't enough.

8

THE 9/11 BONANZA

It wasn't the first time that New York had been attacked. In 1916 German saboteurs blew up an ammunition stockpile, destined to support the Russian war effort, on Black Tom Island, just off the New Jersey shore. The explosion killed four and caused, in 2020 terms, half a billion dollars of damage including to the Statue of Liberty. In 1920 Italian anarchists used a horse and cart to deliver a bomb into bustling Wall Street that killed 30 people. Croatian separatists were suspected to have planted the bomb in LaGuardia airport in 1975 that killed 11. In 1993 the World Trade Centre was attacked by Islamist terrorists using a truck bomb parked in its cavernous basement. It killed six, but failed to achieve its aim of felling the towers. On September 11, 2001, they tried again using aircraft as guided missiles and were spectacularly successful.

The 9/11 attacks were like a far-fetched movie plot. Until then the world's largest terrorist attack was the bombing of an Air India flight over the Atlantic in 1985 that killed all 329 people on board. 9/11 caused almost ten times as many casualties. It struck deep in the heart of one of the world's great cities and, unlike Air India where the only images were of wreckage tossed around in the waves, the sight of planes exploding into the twin towers and the mayhem that followed was indelibly etched into public consciousness. The US, the swaggering global superpower, with its extraordinary military strength, had been stunned by a bunch of zealots armed with box cutters. The world had witnessed many calamities over the centuries, but no single event had the same combination of surprise, body count, destruction, and media coverage. It was the most shocking event in modern human history, and inevitably it brought a monumental response.

The previous chapter described how the attack was carried out and the implications for aviation security. In this chapter, we'll consider the terrorist threat, the US government's response to 9/11, and review its impact on the private security business.

The Home Front

The 9/11 attacks were masterminded by Al Qaeda's head, Osama Bin Laden, from Afghanistan where he was a guest of the Taliban regime. As they refused to present him to the US authorities for trial, within a month of 9/11, the US mounted Operation Enduring Freedom which aimed to drive Al Qaeda out of Afghanistan and bring it to justice. This was followed in 2003 by Operation Iraqi Freedom which in George Bush's words, aimed to "fight them over there, so we don't have to fight them in the US." Both operations certainly proved enduring: US troops were enmeshed in the ensuing wars for 20 years at a cost hundreds of thousands of lives and trillions of dollars, whilst freedom proved elusive.

In the weeks following 9/11, Bush's approval ratings shot up from 50% to over 90%, the nation swung behind him, craving revenge on the preparators and reassurance that similar attacks could not happen again. Reassurance was provided by the National Guard who took to the streets of many US cities. This helped ease people's fears, but it was little more than security theatre that did little to address the risk of further attacks. The problem is that static guards around buildings don't stop attacks by aircraft or car bombs. Effective measures against terrorists are largely unseen: surveillance, forensics, eavesdropping, analysing financial transactions, sharing intelligence with other countries, and targeted arrests once evidence has been gathered. As all this happens backstage, it is unseen by the public, so a visible show of strength by the National Guard was needed to bolster confidence that something was being done.

The public are both the targets of physical attacks and the audience for the accompanying media spectacle. The spectacle causes intense anxiety which translates into political pressure. For these reasons, terrorism is defined as "unlawful use of violence and intimidation, especially against civilians, in the pursuit of political aims." Governments are faced with an unpalatable choice of either giving in to terrorists' demands, or trying to prevent further attacks. There is no easy option. Giving into terrorists' demands haemorrhages political capital and public support, and encourages further extortion. Preventing terrorist attacks haemorrhages money and liberty and is seldom entirely effective. Following 9/11 the US unleashed its military might overseas and introduced a range of anti-terrorist measures on the home front.

Legislation normally takes years to formulate and approve, but the US Government moved with impressive haste and within six weeks of 9/11 pushed through new anti-terrorism legislation called the Patriot Act. The Act created a range of new crimes related to terrorism, established a fund for victims, provided new powers to track and seize terrorist money, tightened up border controls, and, most controversially, made it much easier for the government to intercept communications. The extent of the intercepts was not fully appreciated until it was exposed in 2013 by Edward Snowden, a contractor for the US National Security Agency, who said, "I can't in good conscience allow the US government to destroy privacy, internet freedom and basic liberties for people around the world with this massive surveillance machine they're secretly building."

In the aftermath of 9/11, few were in the mood for discussing liberties and only a single Senator voted against the Patriot Act. There was some comment in the media about how mass surveillance infringed 1st amendment rights to freedom of speech, and 4th amendment rights that prohibit unreasonable searches and seizures. But the prevailing attitude was binary: "if you are not with us, you are against us," and reasoned debate was drowned out by overwhelming emotion. The reality was that in a single stroke Osama Bin Laden had caused the US Government to engage in the mass covert surveillance of its own citizens, eroding the freedoms that are so much part of democracy.

This surveillance regime involved the use of electronic tools to extract data from phone records, bank transfers, flight and hotel reservation systems, emails, and online chats. These were analysed using software such as I2, developed by a British company that was bought by IBM in 2011 for $500 million. I2 creates a visualisation of contacts' networks. If you call your mum, brother, hairdresser, or plumber, their information is shown on screen and linked to you by thin lines, resembling spiders' webs. At a glance you can see who is in touch with who. Everyone's conversations are recorded and analysed automatically. You don't need to follow people with an upturned collar and dark glasses, you don't need to kick down doors. With electronic surveillance you can see peoples' locations, hear what they are saying, review their internet activity, and know how they manage their money. It is an immensely powerful tool. A single intelligence analyst armed with eavesdropping capability is more effective than a regiment of tanks.

Using intelligence to go after attackers might have been the most effective method of countering the terrorist threat, but there was still a huge appetite for a response that could, in the words of President Bush in his Executive Order 13228, coordinate "efforts to detect, prepare for, prevent, protect against, respond to, and recover from terrorist attacks within the United States." The result was the creation of the Department of Homeland Security (DHS) which merged 22 existing agencies into a single organisation with 240,000 staff and an annual budget of $50 billion. This was comparable to the size and budget of the entire British military establishment.

The DHS considered where the next attack might strike. There was a conviction that it wasn't a question of if, but when. As Rudy Giuliani put it, "Anybody—anyone of these security experts—including myself, would have told you on September 11th, 2001, you're looking at dozens and dozens and multi years of attacks like this." The DHS listed 80,000 potential terrorist targets in the US. They included government buildings, malls, chemical plants, museums, theme parks, and schools. Special attention was given to critical national infrastructure: water supply, electrical power, communications, and transport hubs. All places that, if attacked, would have a disproportionate impact on the economy and on public health and safety.

The problem with high-impact and low-frequency events, such as terrorist attacks, is that the likelihood of any one of the 80,000 targets being hit is tiny. The impact is dramatic, but the cost of defending them is enormous. Bin Laden understood this very well, claiming that he aimed at "bleeding America to the point of bankruptcy."

He was encouraged by previous successes. In November 2000, two Al Qaeda terrorists had detonated a small boat full of explosives against the side of the USS Cole, a $1 billion warship, while it was moored in Aden harbour, killing 17 of its crew. The attack cost Al Qaeda perhaps $50,000. 9/11 was of a different order of magnitude. The New York Times estimated that by 2011 it had cost the US $767 billion in damage, economic impact, and homeland security, and a further $2.5 trillion (that's $2,500,000,000) in war costs. The attack cost Al Qaeda $500,000.

To put it another way, for every $1 that Bin Laden spent, the US spent $7 million. But at this emotion-laden time, no one was counting. The government would do whatever was necessary to restore public confidence and to stamp out terrorism. The protective mindset was clearly articulated in November 2001 by Vice President Dick Cheney who said, "If there's a 1% chance that Pakistani scientists are helping al-Qaeda build or develop a nuclear weapon, we have to treat it as a certainty in terms of our response." This was the sort of attitude that reassured the American people.

It was also an approach that was guaranteed to rack up almost limitless expense. The underlying problem is that people conflate risk and fear. Risk is an intellectual exercise in anticipating the likelihood and the impact of future events. Fear is an emotional response to things that frighten you. Emotion trumps intellect every time. In any case, we are equipped to react to predators on the African savannah, not to calculate the odds of dying in a complex modern society. The anxiety generated by the prospect of a terrorist attack has no remedy in statistics.

If we were to scrutinise the data, we would weigh our options more carefully. In their 2011 book, academics John Mueller and Mark Stewart estimated that between 1970 and 2007 Americans had a 1 in 3.5 million chance of being killed in a terrorist attack, a 1 in 950,000 chance of drowning in a bathtub, a 1 in 22,000 chance of being murdered, and a 1 in 8,200 chance of being killed in a road crash. Of course, it's how you die that is important: for politicians, risk isn't a function of impact and likelihood, it's a function of impact and outrage. A single death at the hands of a terrorist will mobilise more media and political attention, and create a greater sense of insecurity, than any number of deaths on the road, or in a bathtub.

There is a concept known as the *value of a statistical life* which is used to estimate the cost of saving a life, or conversely, preventing a death. It is used to evaluate if a policy for, say, road safety, or public health, is cost-effective. The US Government values a human life at around $10 million. That is what it will pay for road engineering at an accident black spot, or the regulation of an environmental hazard. Applying the principle to terrorist threats is more difficult because the data set (the frequency of attacks) is so small. But if we were to divide the home front costs of 9/11 by the number of victims on the day, the expenditure for each was over $250 million. If we include the costs of the wars in Afghanistan, Iraq and the figure is closer to $1billion per death.

There is little doubt that terrorism mobilises a disproportionate level of resources compared to any other risk. Terrorism aligns a wide range of interests: the media for whom terrorist stories grab attention and help to sell newspapers and TV advertising space, politicians who must be seen to protect people, the military and policing establishments who can leverage the terrorist threat to increase their budgets, and the public who have a morbid fear of terrorism's random brutality.

The problem is that you never really know if your counter-terrorist spending is effective. You know what you get for your money if you fund a road, or a hospital, or a water treatment plant. When you pay for counter terrorist activity, you pay for assurance that there will be no attacks. You pay so that nothing happens. But you never know if you are paying too much. And even if there are no attacks, the counter-terrorist establishment will ask for more to maintain its clean sheet.

With the wisdom of two decades of hindsight, we can now see that, although colossal in scale, 9/11 was an isolated attack. Yet at the time it was hard to respond with a calibrated risk-based approach, and rational analysis could not be heard above the fearful chatter. It's therefore worth taking note of the lone voice of former US Colonel Michael A Sheehan who wrote in his 2008 book, *Crush the Cell: How to Defeat Terrorism Without Terrorizing Ourselves*, "we mustn't overreact, it's in our national best interest to simply get over it." He had a point. In the US each year more than 30,000 people are killed in road crashes and more than 45,000 die from gun related injuries. But they don't all occur on the same day, so these deaths do not generate outrage and they are accepted by society with pragmatism rather than moral panic.

From the Basement to the Boardroom

There was no way that the private security business was going to advocate the "get over it" approach. Their clients were fearfully wringing their hands, while their shareholders were quietly rubbing theirs. In the four decades prior to 9/11 the security business had grown steadily, but here was a chance to turbocharge it and the first major opportunity was gifted by the government.

As mentioned earlier, an immediate review of national security prescribed a general uplift in security measures at government facilities, which were mostly guarded by private security, and at critical national infrastructure, 85% privately owned and secured. Nuclear sites alone were allocated an extra $2 billion to fund additional barriers, access control systems, and guards. Most organisations—schools, malls, theatres, supermarkets, office, and residential blocks—also increased the number of visible security officers to reassure staff, students, customers, and residents.

In many cases, the temporary uplift became permanent and increased security became the norm, rather than a short-term response to an emergency. There are two explanations for this: people's sense of anxiety did not subside in the years following 9/11, and with no lowering of the supposed threat, there was no basis on which to reduce security measures, which like a rachet, once tightened are never relaxed. And they don't deliver comfort either. They were either a constant reminder of some unseen threat, or an inconvenience, and they were always expensive, yet we will never know if they were worth it.

Gallop, the US analytics company, takes an annual poll on crime. In the decades before 9/11 it showed that public perceptions of crime broadly aligned with the reality of crime rates, which had fallen steadily during the 1990s. In 2000, only 41% of respondents said that they thought there was more crime in the US than the previous year. After 2001, despite the overall crime rate continuing to fall, the figure jumped to 62% and remained high in subsequent years. This was clear evidence that 9/11 had made people become hyper-alert to security issues and had distorted the way they viewed risk.

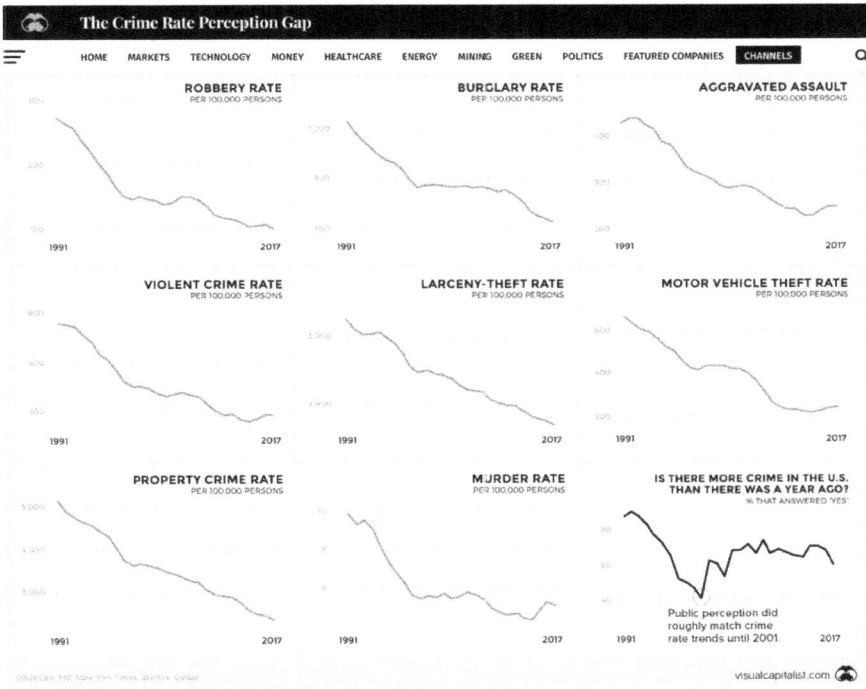

Public perceptions of crime tracked with reality until 9/11.
As Osama Bin Laden said, "America is full of fear ...."
(Courtesy of Visual Capitalist)

All the additional measures were good for the security industry. Because of its sometimes nebulous benefits, security can be hard to sell and it's known as a "grudge purchase." However, after 9/11 the industry had to pedal hard to keep up with demand. Security officers, like traffic wardens, were eyed with little affection or appreciation by the public. The 2009 movie Mall Cop didn't help. The central character, Segway riding, wanna-be cop Paul Blart, becomes a hero in the end, but it is his earnest self-importance that sticks in people's minds. But the status of security officers was elevated after 9/11 as they were enlisted as auxiliaries in the war against terror, and the people who had mocked them were now saying "thank you for your service".

Even the law enforcement community who had traditionally been condescending about private security had an epiphany. The Department of Homeland Security established a network of 78 *Fusion Centers* across the US to share information between the Federal Government, State and local authorities, and included a formal liaison mechanism with private security. Suddenly they had a seat at the table with the police and the FBI. They had hundreds of thousands of officers on the ground, looking out for, not just casual shoplifters, but hardened terrorists. They were well placed for the task; they had training, a formal chain of command, CCTV, and radio communications. If they saw something suspicious, they had a direct line to the Feds, and the Feds could feed them with regular threat updates. Overnight, private had become a valued part of the wider security family.

The call to look out for suspicious activities was widened to include everyone. Advertising executive Allen Kay developed for New York's Metropolitan Transportation Authority, the slogan "see something, say something." It was posted on buses, trains and billboards. The subliminal message was that anyone, with any morsel of information, could foil a terrorist attack that could hit anywhere at any time. In New York in 2002, still reeling from the previous year's attack, 814 suspicious packages were reported. In 2006, after the rollout of the campaign, that number shot up to 37,000. There is no evidence that the campaign resulted in any terrorist plot being uncovered, but it was brilliant at engaging the public, while sustaining their fear and paranoia.

Part of the Business

Following the 9/11 attacks, CEOs across the US asked a range of questions: how likely was it that their business might be the target of a terrorist attack? What would the impact be? How could they account for their staff following an incident? How would they manage a crisis like 9/11? Were business continuity arrangements sufficiently robust? Should all staff be security vetted? Were staff travelling internationally likely to be targeted? From being a grudge purchase to stop criminality, security suddenly became an existential requirement to ensure resilience. Robust security and crisis management mechanisms were seen as business critical and a necessity to maintain the confidence of anxious staff and clients.

Security managers had traditionally been allocated office space in some dismal basement between the cleaners' storeroom and the standby generator. Suddenly they were invited, self-consciously polishing their shoes on the backs of their trouser legs, into oak panelled board rooms to give PowerPoint presentations on the latest threats.

For some, the elevation was permanent. They become fully fledged board members with the title CSO: Chief Security Officer. Others were rebranded as Vice Presidents and reported to a board member. Security became important and influential, and the budgets grew to reflect a change in attitudes from just enough, to just in case. As an American colleague told me, "No one much cared for security before, but it was 'who's your Daddy?' when the shit hit the fan."

Stephen Morrill, Director of corporate security at US technology company Teradyne, put it more politely, "The tragic events of 9/11 remain the single most important turning point in corporate security's new role within business environments both home and abroad." Suddenly security departments were expanded to include a broad range of disciplines: crisis management, business continuity, staff vetting, and travel risk management, along with intelligence cells to monitor events that might impact the business. The role of security had shifted over the years from night watchmen to corporate cop, and was now an all action hero.

The upgrade in capabilities and status was also helped by the collapse of US energy giant Enron as a result of institutionalised fraud. This led to the 2003 Sarbanes-Oxley Act which tightened up corporate governance, accountability, and risk management procedures. A side effect was that major organisations incorporated security into their wider enterprise risk management framework. Security risks were evaluated in the same way as risks to supply chains, compliance, or market share. It heightened awareness that a serious security incident could impact operations, staff motivation, brand reputation, and ultimately share price. Security became more accountable, more metrics-driven, with its own set of performance indicators, just like any other department.

9/11 also further tightened the sinews between the insurance business and the security business. The insurance business is like a casino: the house always wins. The problem is that terrorism, being a low-frequency but high-impact event, makes the risk, and therefore the premium, hard to calculate. Insurance pay-outs for losses resulting from the attacks were in the region of $40 billion. As they couldn't guarantee a win, the insurance businesses' immediate response was to withdraw cover for terrorist attacks.

Lack of insurance for a catastrophic risk was a predicament for the wider economy. Investors would not fund a business if it was not insured against terrorist attacks—which in the post-9/11 period seemed certain to happen again. The issue was solved using a template established by the UK in 1993 following

IRA bombing attacks in The City where the Government agreed to cover insurance company losses. This meant that insurance companies could issue policies secure in the knowledge that their exposure was limited, and whatever the spin of the wheel, the house would not lose.

Security and insurance had long been partners in preventing loss. The more effective security is, the less likely that insurance will have to pay for losses including stolen goods, damaged property, business interruption, negligence, ransoms for kidnapped executives, traumatised staff, or litigious customers. Security is insurance's best friend, its tame rottweiler that keeps bad things from happening and protects profit.

Following 9/11, once the government had agreed to underwrite insurance company losses, there was massive demand for insurance. A further condition for organisations wishing to take out terrorist insurance was additional security measures to reduce the likelihood of an attack, and effective crisis management and business continuity plans to lessen the impact should an attack occur. To make it easy, many insurance companies partnered with security companies that could provide the full range of services to clients. Everyone was a winner: the insurance company was less likely to have to pay out, the security company received a retainer and access to consulting opportunities, and the company taking out the insurance has an upgraded crisis response capability and was better protected.

The private security sector had developed gradually prior to 9/11 by a steady 5% a year. However, afterwards, like a teenager experiencing a growth spurt and jumping a shoe size every few months, it ramped up to 6.5% a year. It wasn't a get-rich-quick option, but it was a sound investment and a safe long-term bet as levels of public anxiety remained high.

It attracted the attention of big money: banks, private equity, and hedge funds who recognised that capable, multi-disciplinary, private security companies could help to protect their major investments including oil and gas, mining, aviation, infrastructure, and large corporations. In a complex globalised world, the effective management of security risk was seen as a business enabler, a source of competitive advantage, an aspect of social responsibility, and a guarantor of resilience in uncertain times.

Bring on the Tech

Few could forget the heart-rending missing persons notices that were taped to billboards, trees, and bus shelters in the days after 9/11. It was an immensely emotional odyssey. Were loved ones crushed beneath the towers, lying in hospital, just out of town, or out of touch? The first thing that any CEO would ask after an attack in any part of the world was, "are our people ok?" There would be apprehension until everyone was located.

Traditionally, when there was an incident, people were accounted for using a roll call, or headcount, like a school assembly. In the 1990s all that changed, with the arrival of personal mobile phones it became possible to contact people using a phone tree. No more setting eyes on people, you could just call them. An organisation would maintain lists of staff who would be contacted using a cascaded hierarchy until everyone had been checked off. It sounds quick and simple, but the reality can be different. I once initiated a phone tree to account for 50 staff in a European city following a terrorist attack. We contacted 40 within a couple of hours, but it took 4 anxiety filled days to contact the remainder who had their phones off, were out of battery power, or out of range, or didn't recognise the caller number and wouldn't pick up. The final person to be ticked off the list, was found prostrate, not in hospital, nor in a morgue, but on a sun lounger in the Maldives.

Organisations then tried to account for people by using building access control and corporate travel booking databases. But the big breakthrough arrived with the introduction of smart phones in 2007 which were able to track people using the phone's GPS (global positioning systems). Larger organisations started using automated messages services developed by companies such as Everbridge and Vismo, which used advanced IT systems to monitor the internet for breaking news about security incidents, check that staff were ok and issue real-time updates. Instead of hours or days, they provided an instant head count along with situational awareness.

It wasn't just organisations that wanted to account for people. Anyone caught in the vicinity of a newsworthy tragedy sought to reassure loved ones that they were ok, or more likely, the other way round, as parents would see an incident on the news and immediately call to check on their kids.

In 2014 Facebook harnessed this demand in a feature known as *safety check*. Facebook tracks users' location and, if there is an incident, it immediately sends those in the vicinity a notification asking if they are ok and would like to inform their contacts. No need for all those "don't worry Mum" calls, just check Facebook and she'll know that you'll be home for dinner.

Technology reset expectations of the speed of accounting for people. In the old headcount days we were sanguine if people could not be contacted immediately. But technology gave us instant communications, which fuelled our urgency to check on people. Anxiety accumulates with each passing minute, everyone assuming the worst, until each is found and ticked off the list. The tools designed to calm us have the opposite effect.

Until the 2000s organisations may have had some form of security control room with a red eyed person at a grubby desk surrounded by a phone, an ashtray, a radio, a whiteboard with scribbled lists of emergency numbers, a kettle, a microwave oven, the smell of the last meal lingering in the air, and a CCTV monitor that attracted fewer glances than Sports Illustrated calendar next to it.

The post-9/11 world created new demands and expectations, and required new capabilities, funding, and technology for security control rooms. The upgraded versions were rebranded as the SOC—security operations centre—and they resembled the flight deck of the Starship Enterprise with rows of earnest staff peering at large computer monitors. The whiteboards were replaced by flat screens showing news feeds, electronic maps of company sites, geopolitical risks, security alerts, and the exact locations of patrolling security officers with a feed to their body cams. Telephones trees had been replaced by mass text or email messaging. CCTV was monitored by artificial intelligence that alerted SOC staff to intruders or unattended baggage. Automated systems identified who was in which building and pinpointed the location of travelling staff.

High tech SOC, mission control for a modern organisation.
(Courtesy of Everbridge)

The SOC became a high-tech hub, a mission control centre for the modern organisation. From here everything and everyone could be tracked, monitored, and contacted. If there was another 9/11 the SOC could immediately assess the damage, mobilise resources, liaise with the authorities, account for everyone, pass information to staff, and initiate crisis and business continuity plans. The SOC had it nailed.

In the 20 years following 9/11 the value of the US private security market doubled from $15 billion to more than $30 billion, while the number of people working within it increased from 900,00 to 1.1 million. The value of the market doubling while the headcount nudged up only 10% reflected the automation of traditionally labour-intensive processes, such as access control and CCTV surveillance. Warm blooded staff were replaced by cool new machines, at a fraction of the cost. At the same time, remaining security staff became more professional, better trained, and there were more executive, management, and specialist positions. Security had become bigger and better, spurred by a terrible event which hard wired people to always consider worst case scenarios. It's no surprise that security professionals are fond of repeating Winston Churchill's famous line, "never let a crisis go to waste."

9

7/7 PERSPECTIVE LOST

The shock waves of 9/11 quickly surged across the Atlantic. If the US, the great Satan, could be attacked, then the little Satan, the UK, must surely be a target too. In many ways, the UK was better prepared for attacks than the US, and it certainly had more experience of civilian casualties. In the century prior to 9/11, 51 New Yorkers had been killed in four attacks. In the same period, tens of thousands of Londoners had been killed in hundreds of attacks. In WWI 667 Londoners were killed in German bombing raids. In WWII 30,000 Londoners were killed during the Blitz. Throughout the 1970s, 80s and 90s, during the low-intensity conflict known as the Troubles, the IRA routinely detonated bombs across the capital. They caused immense disruption and killed 50 people. Their last major bomb, in 1996 in the Docklands business district, killed two and caused £150 million worth of property damage. Having experienced so much carnage, over so many years, the UK had developed a doughty stoicism known as the Blitz Spirit and the motto "Keep calm and carry on".

The UK's critical national infrastructure, along with many government and military sites, were given substantial protection to mitigate against the IRA attacks which included car bombs, letter bombs, incendiary bombs, time-delayed bombs, suitcase bombs, and even improvised mortar bombs. The 1998 Good Friday agreement brought much of the violence to an end, and, after 9/11, the intelligence and security establishment quickly shifted its focus from the Irish to the anticipated Islamist terrorist threat.

The IRA saw the UK authorities as legitimate targets and while aiming to disrupt transport and businesses, they generally avoided killing civilians on the

British mainland as it undermined their cause, even amongst Irish nationalists. Furthermore, IRA terrorists didn't want to get caught or identified, so CCTV (described in chapter 6) had some deterrent effect on their operations. They didn't want to get killed either, preferring to live to fight another day, so they never went in for suicide attacks. They often telephoned warnings before their bombings to reduce casualties. And it was clear what they wanted: a united Ireland. So, there was a rationale and a degree of predictability to their attacks.

Islamist terrorists by contrast sought to kill large numbers of civilians and they *wanted* to die in the process. They appeared more driven by hate than political purpose. Success for them was a high body count, which generated media coverage, which amplified the public's fear and outrage. IRA terrorism had been met with a degree of weary pragmatism. But on 9/11, 3,000 people were killed on a single day by Islamist terrorists, almost as many as in the 30 years of the Troubles. Weary pragmatism was replaced by deep anxiety.

Shocking but not Surprising

The 9/11 attacks seemed to come out of nowhere. They shattered Americans' sense of security so deeply that 20 years later it remains broken, perhaps irreparably. By contrast, London's 7/7 attacks in 2005 were shocking, but not surprising. By then the UK had joined the war on terror and was fighting alongside US forces in Afghanistan and Iraq. Islamist terrorists had bombed a nightclub in Bali in 2002 (with 23 British citizens amongst the 202 dead), the British Consulate in Istanbul in 2003 (killing 16 including 10 Consulate staff), and commuter trains in Madrid in 2004 (killing 193, although no British citizens were amongst them). It seemed only a matter of time before a major Islamist terrorist attack on British soil.

On 7/7, four suicide bombers targeted London's transport network almost simultaneously. Three detonated on underground trains, the fourth detonated on a double-decker bus. In total 56 people were killed. Apart from the blowing up of the Pan-Am flight over Lockerbie in 1988, it was, by some margin, the worst terrorist attack in the UK. But it wasn't the largest loss of British lives in a terrorist attack. That grim record went to the 9/11 attacks, where 67 of the 3,000 killed were British citizens.

Preventing another 9/11 was fairly straightforward. Identify and profile all passengers in advance to see if they pose a risk, search them and their luggage before boarding, and harden cockpit doors (see chapter 7). Preventing another 7/7 was a different story. London's transport is inherently vulnerable to terrorists. There are more than 10 million passenger journeys on London's tube and bus network every day. There is no identity check, and it is impractical to set up airport-style security at every tube station and bus stop. The extensive CCTV coverage across the capital is no deterrent to suicide bombers. Armed guards on London's 8,000 buses and 500 tube trains would be ineffective and prohibitively expensive. How to solve this conundrum?

Drawing upon decades of counter-terrorist experience the government developed a strategy known as CONTEST. This had four elements: Prevent, Pursue, Protect, and Prepare.

- **Prevent** people from being radicalised by challenging those who promote extremism.
- **Pursue** potential terrorists using surveillance and intelligence, and put them on trial.
- **Protect** potential targets through physical security measures.
- **Prepare** responses to any attack.

Of these, there was compelling evidence that the pursue element—surveillance and intelligence—was the most effective. Between 2001 and 2021, only a handful of attacks materialised while 1,004 people were convicted for terrorist-related offences. Some were in the advanced stages of planning attacks, although most had only accessed terrorist material online or provided money to overseas terrorist organisations.

Be Alert

A key player in the intelligence efforts was JTAC, the Joint Terrorist Analysis Centre, formed in 2003 and housed behind the stern, Portland stone façade of the MI5 headquarters in Westminster. As well as analysing the many nuggets of information that built up a picture of potential terrorist activity, it had the unenviable task of providing warnings of future attacks. JTAC published a nationwide terrorist alert framework with five levels of threat ranging from *critical*—an attack is imminent, to *low*—an attack is unlikely.

"Threat levels," says MI5's website, "are a tool for security practitioners ... and the police to use in determining what protective security response may be required." Each level triggers a sequence of security measures at potential targets: frequency of patrols, protocols for searching visitors, numbers of guards, and rehearsals for contingencies. Maintaining a heightened security posture is hard work, resource intensive, and intrusive. It's like trying to sprint in a marathon; you soon get tired. Varying the threat levels allows performance to be sustained and more efficient use of resources.

Since they were introduced in 2006, the UK's threat level has never been below Substantial which means: *an attack is likely*. Most of the time the level has been at Severe: *an attack is highly likely*, and on four brief occasions it has risen to Critical: *an attack is highly likely in the near future*.

Changes in the threat level often result in verbal gymnastics, such as this from Home Secretary Alan Johnson in January 2010, "JTAC has today raised the threat to the UK from international terrorism from Substantial to Severe. This means that a terrorist attack is highly likely, but I should stress that there is no intelligence to suggest that an attack is imminent."

The problem is that threat levels provide no indication of where, or when, an attack might occur, or what form it might take. It baffles the public who puzzle over the difference between an attack being "likely" or "highly likely". The key purpose is to demonstrate that the government is monitoring the situation. If there is an attack, JTAC could say it had pointed to the prospect of something happening. If no attack occurs, it could say that the likelihood of one remains real.

The source of power for JTAC and all intelligence organisations and their spooks, as they are often known, is making people believe that they know something that you don't. If they do know something they can't tell you. And if they don't know something they wouldn't tell you. Spooks realize that you have seen the Bourne Supremacy, and that whatever you imagine they know is much more exciting that what they do know. They want you to believe that they possess some dark alchemy as they fix you with their sphinx-like smiles.

The reality is that most intelligence work involves trawling through online data bases, rather than meeting shadowy figures in the souks of Aleppo. And if JTAC knew of an impending attack, it would pass on clear, actionable information, rather than a vague general alert that suggests an imminent threat. JTAC understands what American polymath Tom Lehrer meant when he said, "always predict the worst and you'll be hailed as a prophet."

Politicians recognise that the frequency of attacks is very low but the political consequences, of even a small incident, are very high. As David Cameron said in 2015, "Protecting the British people is my number one duty as Prime Minister." He was also protecting his own position. As journalist Simon Jenkins argued in The Spectator in 2016, "To politicians, it [terrorism] is an opportunity to flex muscles, brandish guns, boast revenge. Counter-terror theory may advise caution and an emphasis on normality. Political necessity counsels the opposite; the trumpets and drums of battle. It requires the terrorist's deeds to be amplified, headlined, exaggerated to justify a warlike response."

As a result, politicians respond to public fear rather than actual risk. Professor David Nutt, the government's chief drug adviser, discovered this the hard way when he was sacked in 2009 for pointing out that the drug ecstasy was statistically no more dangerous than horse-riding. Given Professor Nutt's experience few were willing to put the terrorist threat into perspective.

Between 2000 and 2020, 88 people were killed by terrorists in the UK, 82 of them in just three attacks (7/7, Borough Market and the Manchester Arena). During that same period, around 16,000 people, or 800 a year, were killed by murderers. So, terrorists are responsible for roughly four deaths a year, or about 0.5% of all killings.

The table below shows the average number of people who die prematurely each year in the UK by a variety of means.

Suicide	6,000
Falling	4,000
Car crash	1,700
Murder	800
Drowning	400
Cows	6
Terrorists	4
Dogs	4
Horses	4
Lightning	2

It's clear that the chance of being killed by a terrorist is in the same range as being killed by a horse, a dog, a cow, or by lightning. To add further perspectives, around 160,000 in the UK die of heart disease annually. So, you are 35,000 more likely to die because of your life-style choices than in a terrorist attack. And since its inception in 1994 there have been more than 6,600 lottery millionaires, so you are more than 50 times more likely to become a lottery millionaire than a terrorist fatality.

Both terrorists and lottery operators know that the public is poor at understanding probabilities. Canadian-American psychologist Steven Pinker describes terrorism as "a unique hazard because it combines major dread with minor harm." The National Lottery in the UK used the advertising slogan, "It could be you" to fuel our hopes. Terrorists fuel our fear of random death in the same way. Both feed our biases and make unlikely possibilities feel like imminent probabilities.

Shoot the Messenger

"The purpose of terrorism lies not just in the violent act itself. It is in producing terror," said Tony Blair in 2003. Terrorists are responsible for the violence, but the fear of terror is amplified by the media. As American historian Walter Laqueur wrote in 2000, "It has been said that journalists are terrorists' best friends, because they are willing to give terrorist operations maximum exposure. This is not to say that journalists as a group are sympathetic to terrorists…. It simply means that violence is news, whereas peace and harmony are not. The terrorists need the media, and the media find in terrorism all the ingredients of an exciting story."

The media is the weak spot in any counter-terrorist strategy. The media is a rage farm that cannot easily be controlled by the government in democratic countries. Yet it can be exploited by terrorists, as Yuval Noah Harari wrote in the Guardian in 2005, "Terrorists don't think like army generals; they think like theatre producers." The more sensational the spectacle, the greater the audience, and the larger the profits for media outlets. There is a symbiotic relationship between terrorism and the media. This dynamic would be calmed, and terrorists denied the oxygen they seek, if the media was prevented from giving prominent coverage to terror-related news.

In June 2017 terrorists used a vehicle and knives to attack people on London Bridge and at nearby Borough Market, with knives, killing 8. There was massive media coverage in the weeks that followed and the market was closed for 12 days. By contrast, in Istanbul in November 2022 terrorists detonated a bomb in a busy shopping street killing 6. Within an hour of the attack, the Istanbul Criminal Court issued a blackout of broadcast, audio, and social media news about the incident. The scene of the attack was quickly cleared up, thousands attended a ceremony to comemorate the dead the following day, and the street was immediately opened for business. This demonstrates that it is possible to take a pragmatic approach to terrorism that aims to restore normality and public confidence as quickly as possible. Preventing the media from running away with an emotionally engaging narrative is key.

Sun Tzu, the ancient Chinese military strategist, summarised the art of psychological warfare as 'kill one, frighten 10,000.' He was right. In the decade following the 7/7 attacks, only a single person in the UK was killed by terrorists (the unfortunate off-duty soldier, Lee Rigby, in 2013). However, surveys by YouGov showed that the proportion of people who thought that the threat of terrorism had increased jumped from 25% in 2010, to 74% in 2016. There had been attacks in other countries, and the British military was fighting in Afghanistan and Iraq, but the survey result further demonstrates a clear disconnect between the reality and the perception of the terrorist threat.

The government, too, struggles to find the right balance between enlisting public support for action against terrorists and generating suspicion and unease. In November 2016 the Minister responsible for railways, Paul Maynard, announced the launch of the "See it, Say it, Sorted" campaign, despite no terrorist attacks occurring on the transport network since the 7/7 attacks 11 years previously. "See it, Say it, Sorted" was a copy of the US "see something, say something" campaign which encouraged people to report suspicious activity to the police. The message was relayed using 11,000 posters and more than 30,000 loudspeakers on railway trains and stations.

I became intimate with the campaign during my journey to work in central London which involved eleven stops on a commuter train, followed by seven stops on the underground. The pre-recorded message was delivered at every station. In an average week, I heard the message 180 times and saw it on at least as many posters. After weeks of messaging, I experienced a strange cocktail of paranoia, anxiety, and deep irritation, as the "See it, Say, Sorted" earworm turned in my head.

Paul Maynard said: "We want to send a clear message to anyone threatening the security of the rail network that there are thousands of pairs of eyes and ears ready to report any potential threat." On one level it was successful: there was a 365% uptick in the number of people *seeing it* and *saying it* to the police. Yet, in common with the US campaign, not a single call was held up as successfully thwarting terrorist activity.

Along with paranoia, the other thing that increased was the UK's annual counter-terrorist budget. Ben Wallace, Minister for Security and Economic Crime at the time, announced in parliament that "In 2015, the Government increased counter-terrorism funding by 30%, from £11.7 billion to more than £15 billion." He was referring to a five-year period, so the annual

132 THE RISE OF SECURITY and Why We Always Want More

Providing reassurance, or stoking anxiety?

counter-terrorist budget was £3 billion. By comparison, during the same period, there was much controversy about the £3 billion cost of the UK's aircraft carrier HMS Elizabeth. Especially as there were insufficient funds for the accompanying aircraft and supply ships.

It seems that the UK spends more on countering a handful of fanatics armed with kitchen knives and homemade explosives, than on projecting British power overseas. By inflating perceptions of the risk of terrorism we end up allocating disproportionate financial resources that could be better spent elsewhere.

Troops Against Terror

In 2017, reality briefly caught up with perceptions. During that year in the UK, 35 people were killed in four unrelated terrorist attacks. Three of the attacks were in London (Westminster, London Bridge and Finsbury Park), but the most serious, and the first suicide bombing since 7/7, was outside the Manchester Arena following a concert by American singer Ariana Grande, which killed 22 young people. This was the sort of attack—the bombing of a crowded space—that had been feared for more than a decade.

The government immediately ordered armed troops to Manchester to assist the police with protective security duties. It was a controversial decision. The last time that the army had been called upon to make a show of force on the streets of mainland Britain in peacetime was in 2003 when Tony Blair sent tanks to Heathrow Airport following a threat to UK aviation. Some people found this reassuring, but it was derided by others who pointed out that it was pure "security theatre" as tanks provide no protection against suicide bombers on aircraft. For the travel sector, it was, said Tom Jenkins, Head of the European Tour Operators Association, "a public relations disaster" as tourism from the US that year dropped by 80%.

Putting troops on the streets to counter terrorism creates a wartime atmosphere, burns up resources, but does little to address the terrorist threat. It also creates a bureaucratic predicament: removing the troops requires the intelligence services to declare the threat sufficiently reduced to return to normality. However, the prospect of an attack can never be entirely eliminated. So, once troops are deployed, it's easier to maintain them and risk criticism for wasting resources, than it is to withdraw them and risk criticism if there is an attack, even if they would have been ineffective at stopping it anyway.

Cool heads prevailed and troops were quietly withdrawn from Manchester's streets after a week. In France, it was a different story. In 2015 following a terrorist attack in Paris, 10,000 troops in green camouflage, carrying assault rifles, were deployed to government offices, transport hubs, and shopping centres across the country. There was little evidence of their effectiveness, or that they calmed public anxiety, but on at least half a dozen occasions they were themselves targeted by extremists, venting their rage on what they saw as symbols of repression. Seven years later the troops remain on the streets and look set to become as much a part of French life as baguettes and boules.

Protect Flaw

One of those killed by the suicide bomber at the Manchester Arena was 29-year-old Martyn Hett. Following the attack, his mother, Figen Murray, campaigned for a Protect Law (also known as Martyn's Law) which aimed at making venues safer from terrorist attack saying, "It's crucial this law is brought in and applied to all public venues because protecting the public from terror attacks is a priority." The proposed law would oblige venue operators to institute a range of counter-terrorist measures including security training for staff, crisis action plans, risks assessments and the mitigation of risks through security checks for visitors, hostile vehicle mitigation, safe rooms, CCTV, and the deployment of additional security staff.

Figen Murray spoke emotionally and convincingly about the need to protect people from attack. "It's too late for me," she said, "I've lost my child—but I'm committed to doing all I can to stop other families having to go through the nightmare that we are." The solution, she said, in an interview with The Sun newspaper in December 2020, was easy, "I bought six handheld metal detectors for less than £130 online, that's under £22 each, to show just how simple it is." She said that "Martyn's Law is a simple set of measures that are proportionate. A small cafe wouldn't need the same security measures as a large stadium or arena".

The problem is that no "simple set of measures" will stop a suicide attack. If security staff, waving a metal detector wand, were to identify a bomber trying to enter a building, he is likely to detonate on discovery. More likely he would avoid the security check and detonate amongst crowds outside, at a bus stop, in the shop next door, or in the town square. A suicide bomber is like a missile: once launched, he must detonate somewhere, and no amount of metal detectors, CCTV or security guards are going to stop him.

There are 650,000 public venues to which this law could be applied, from sports arenas to shopping arcades, bars, beaches, concert halls and churches. For small venues with a capacity of 100, the cost might be as little as £500 a year (for a risk assessment, paying staff to be trained, CCTV, access control, security staff etc.). For large venues, it might be as much as a million a year. But effective measures against a terrorist that might use a bomb, a vehicle, a knife or a gun, rather than security theatre, would cost much more. The starting price for modest upgrades at all the 650,000 venues is likely to be over a £1 billion, in addition to the government's £3 billion counter terrorist budget. It would be a further boost to the UK's private security business, which doubled in size between 2000 and 2020 to reach 400,000 staff and a turnover exceeding $6 billion. But would it make anyone any safer?

Had the law been in place for the past 20 years there is scant evidence that it would have saved a single life, including at the Manchester Arena where the bomber detonated outside the venue.

Stopping a determined terrorist is no easy task. In 1990 the Metropolitan Police provided security for a conference in London to discuss the latest wave of IRA terrorism, that attracted senior figures from the UK's counter-terrorist establishment. Despite comprehensive security measures being in place, the IRA succeeded in placing a bomb inside the lectern where speakers, including the Metropolitan Police Commissioner, Sir Peter Imbert, were due to address the delegates. Fortunately, it was spotted by a technician as he checked the microphone. The venue was evacuated, and the device was disarmed. Lives were spared, blushes were not. But the episode demonstrated how terrorists can evade the most thorough of security measures.

In the early 1990s, I was part of an Army Counter-Terrorist Search Team that supported the police. The team had been set up in 1984 after an IRA bomb, fitted with a long-delay timing device from a video recorder, detonated in the Grand Hotel in Brighton where many members of the Conservative Party were staying during their annual conference. Five people were killed, but the main target, Prime Minister Margaret Thatcher, narrowly escaped.

Our task was to search for bombs ahead of events attended by the highest levels of government. Searches would be planned weeks in advance. All venue staff would be vetted and issued with ID passes for the event. A police cordon would be placed around the venue, while our team picked through it with specialist equipment designed to detect bomb components: explosives, electronic timers, batteries, and wiring. We examined every chair, cupboard, wall, roof void, floor space, table, book, coat rack, desk, radiator, and curtain. We dismantled, examined, and reassembled everything.

The cordon remained in place for the event, with attendees passing through airport-style screening before taking their seats. A thousand-person venue would take a team of twenty specialists 48 hours to search, and the cordon and access control involved hundreds of police officers. This is how to protect venues from terrorist attacks.

Our nightmare scenario was a suicide bombing. We knew that there was no easy solution. We studied Israeli checkpoints designed to counter the threat. They used blast-resistant chambers that allowed a single person to enter, observed by CCTV, with instructions barked through a crackling intercom by security staff. "Show your ID to the camera.… why are you here.… take off your clothes … turn around … pass your clothes through the X-Ray …." Inside the chamber, there was a machine to detect explosive vapours and a robotic arm to restrain the person remotely should they be suspect. It was slow, intrusive, and very expensive. But it was the only effective protective security measure against a determined suicide bomber.

People might grimly surrender to airport security with its queues and shoe-less shuffles through metal detector arches, on their way to a Mediterranean beach. But they would soon become impatient if they experienced similar security on daily visits to a supermarket, cinema, gym, or at a wedding venue. And they certainly wouldn't tolerate being stripped naked in a blast-proof chamber.

In the public's mind, and often in reality, terrorists are from ethnic minorities. Keeping an eye out for terrorists would inevitably mean being very suspicious of people with dark skin, especially if they are carrying a rucksack or wearing a hoodie. One of the aims of terrorism is to force authorities to institute disproportionate measures that end up eroding their moral authority and budget, while alienating minority groups. An unintended consequence of excessive security measures is that they divide communities, inconvenience everyone, and may actually increase the threat that they aim to counter.

However, Figen Murray's campaign moved a step closer in June 2021 when Sir John Saunders, Chair of the Public Inquiry into the Manchester Arena attack, issued his report. He said that, "inadequate attention was paid to the national level of the terrorist threat by those directly concerned with security at the Arena. The threat level was severe. That meant that a terrorist attack was highly likely."

This exposes a weakness with the government's terrorist threat levels. They institutionalise paranoia and fixate the public on a single remote risk. In fact, no concert hall in the UK had ever been attacked by terrorists. The odds of any of the 650,000 venues, including the Manchester Arena, being attacked on any given day were extraordinarily small, and no one had been killed by a terrorist bomb in the UK in the previous 12 years.

But Saunders was clear, "Everybody concerned with security at the Arena should have been doing their job in the knowledge that a terrorist attack might occur on that night."

Saunders' view represents a shift in emphasis. Rather than the government having prime responsibility for protecting the public from terrorists, the load is now placed on venue owners. The notion of owners having responsibility for securing their own private property is long established, and it works well for crime and safety. But terrorism is a political act that requires security measures that are well beyond the ability of owners. If a terrorist attack was "highly likely" at Manchester, there should have been armed police on duty that night, rather than leaving it to civilian security staff to face a suicide bomber.

JTAC's vague threat alerts should not be used to force venue owners into implementing expensive and ineffective measures to prevent a terrorist attack. Alerts should only be issued where there is specific intelligence that can be acted upon by properly equipped authorities.

Another unintended consequence of the Protect Law is that it would be applied unevenly by venue owners. Some would take it seriously, others would be more reluctant, and all would be conscious of the resource implications. There would need to be an inspection regime and penalties for owners who fail to meet the standards. If only 1% of the 650,000 venues fail to meet the standards each year, more people would be convicted of breaching the Protect Law than of breaching the Terrorism Act.

The inspection regime would also be resource intensive. If a single inspector could review 120 venues a year, 5,400 would be needed to make annual inspections, or 2,700 if they were every two years. Take the smaller figure and add a further 300 management, administration, training, and HR staff, and at 3,000 strong the inspection organisation would be the size of the British Transport Police and would likely cost £150 million a year. Some venue owners might welcome its, others would rail against the additional measures they would inevitably be forced to implement.

The Protect Law would create a wider conundrum. If there was an attack at a venue that did not fall under its remit, such as a newsagent, or a hairdressers, or a doctor's surgery, would you then extend the Protect Law to include them, or would you recognise the futility of trying to protect everywhere from unlikely occurrences? As Dwight Eisenhower said, "We will bankrupt ourselves in the vain search for absolute security."

10

GUNS FOR HIRE

There's no sign on the door. Nothing to distinguish the building from others in the elegant Victorian red brick terrace just a stone's throw from Harrods, in London's exclusive Knightsbridge neighbourhood. You press the video intercom, and a voice quickly responds. "Can I help you?" If you are expected, the heavy black lacquered door clicks open and you enter the hall, sign in at reception, and hang your coat on the rack. You walk up the thickly carpeted stairs flanked by black and white photos of handsome men and women. They could be movie stars, but they are members of the Special Operations Executive who were killed in action. This is one of the most select establishments in London: The Special Forces Club.

The lounge is on the first floor. Its walls are covered with oil paintings of military scenes: determined-looking soldiers, low-flying aircraft, and, bizarrely, the Queen Mother with a corgi. At the polished oak bar, there is a picture of a parachutist with the words "if you are dropping in, introduce yourself to the person on your right." This is the heart of the club where people "pull up a sandbag and swing the lamp." Old mates are remembered, stories are swapped, and deals are made on a nod and a handshake. It is also the spiritual home of Private Military Companies (PMCs).

PMCs are gunslingers at the high end of the private security business. They operate internationally, often in places where bullets are likely to fly: conflict zones, failed states, and countries where the government does not have a monopoly on violence. PMCs recruit competent ex-soldiers who keep a low profile and are comfortable working off the beaten track. To some, they are hired guns, soldiers of fortune, dogs of war, mercenaries. To others, they are simply skilled contractors working for governments or multinational corporations in high-risk environments.

Dogs of War

The term mercenary comes from the Latin *mercenarius* which means "hired for money." It has pejorative, somewhat sinister connotations, and a lot of historical baggage. But why should soldiering for money be any less worthy than being a carpenter or a pilot? Even the Vatican hires mercenaries. The Swiss Guard have been protecting the Pope since the 16th century. With their blue and orange baggy pantaloons and their oversized berets, they may not look especially fearsome now, but back in the day they were the hard men of Europe, famed for their courage and invincibility. And their loyalty was never in doubt, as long as they were paid.

Other, less godly people, were prepared to pay others to fight for them too. Mercenary units fought for the Romans, the Normans, and for the Italian city-states. And, arguably, depending on how you define mercenary, some are still in service: the French and Spanish both have Foreign Legions, and the British continue to recruit Gurkhas from Nepal. But in the 20th century mercenaries became more controversial. As the tide of colonialism ebbed, a new breed entered the guns for hire business.

Most people enjoy life in the military: the camaraderie, the adventures, the physicality, the code of honour, the sense of purpose, and the patriotism. Those who serve a full military career are retired at 40 if they are enlisted, or 55 if they are commissioned officers. So, they leave knowing their best years are behind them, but they still have a bit of snap in their celery. Adjusting to civilian life can be a challenge. Most transition well, but others are more restless and struggle to find their place. They miss soldiering and they crave more adventure. Being a hired gun provides a sense of purpose and one last chance to work with brothers in arms.

And then of course there's the money. You can get by on a military pension, but only just. PMC work can be lucrative, although often it is mostly short-term contracts with long stretches of not much in between. Being on a succession of contracts is known as *The Circuit* and the saying is, "there's no security, in security." Or as one of the disgruntled old guard told me, "you're like a monkey up a bloody tree, waiting around for someone to throw you a few peanuts."

The modern guns-for-hire business has two broad strands, although to the public, little separates them. But inside the business, there is a clear distinction between those that are paid to provide protective services, and those who are paid to fight offensive operations. The former guard facilities such as pipelines and embassies, the latter are your classic mercenaries, paid to actively fight. The highest profile modern mercenary organisation is the Russian Wagner Group which has shadowy operations in Ukraine, Syria, Mali and elsewhere.

The modern mercenary business traces its heritage back to "Mad" Mike Hoare who was commissioned into the British Army in WWII, serving in Burma and India, and leaving as Major in 1945. He married that same year, had three children, and seemed set for a quiet life after qualifying as a chartered accountant. But longing for adventure he moved to South Africa in the 1950s, and by the 1960s he decided he would rather count bullets than other people's money and set off for the Congo to fight for the government against communist rebels. Hoare raised a small army of fellow WWII veterans of various nationalities. Apart from fit men with fighting skills, he sought what he called, "tremendous romantics." For him, as it was for many hired guns, it was mostly a great game. But he played ruthlessly, claiming that his unit had killed more than 5,000 rebels.

Hoare was the technical advisor on the film The Wild Geese which was loosely based on his exploits. Starring Roger Moore and Richard Burton, it was one of the biggest hits of 1978, released with a royal premiere attended by the Duchess of Kent. The plot starts with a smooth London banker hiring a retired British Army officer to raise a small force to rescue an African President.

The film was well received at the time and it embedded mercenaries in popular imagination as lost, yet noble souls, seeking one last throw of the dice. Today it would be judged more on its body count and casual racism, and we would struggle to imagine a film celebrating mercenary exploits in former colonies receiving royal endorsement.

Perhaps swelled by the movie's success, Hoare decided on another adventure: leading a coup in the Seychelles in 1981. Posing as a drinking club "Ye Ancient Order of the Frothblowers," Hoare and his team of 54 mercenaries landed in the Seychelles where they were rumbled by an alert customs officer at the airport. A firefight broke out ending the coup attempt and they forced an Air India plane to take them to South Africa. Hoare and his "Frothblowers" were arrested on arrival, convicted of hijacking, and sent to prison. Hoare also had the distinction of being the only member of the British Institute of Chartered Accountants ever to be expelled for mounting a coup.

Bob Denard, the similarly notorious French mercenary, also fought in Congo after leaving the French Navy where he had served in the Algerian War. He went on to fight in Yemen, Rhodesia, Angola, Iran, and Nigeria. He might have remained known only to a select circle if he too hadn't had such a fondness for mounting coups in another African island country: The Comoros. Four times. A colourful character, Denard was jailed twice, changed religion four times (from Catholicism to Judaism, to Islam, and back to Catholicism) and was married seven times.

Both Hoare and Denard were men of their time, chancers plying their trade in the fading embers of empire. Lost souls whose love of soldiering was blended with uncompromising masculinity and unrefined morals. Recognising that there were many such spirits, the self-titled, "misfit Vietnam vet" Lt Colonel Robert Brown published Soldier of Fortune magazine in 1975. It featured articles about military adventures, promoted first and second amendment rights (rights to free speech and to carry arms) and carried advertisements for guns and gear. It is still published online and has a firm following amongst right-wing gunslingers. It also appeals to a strange category known as "Walters," normally civilians, named after the James Thurber character Walter Mitty, who was a hero in his own fantasy world.

Hoare and Denard were undoubtedly rogues with blood on their hands. Yet their exploits received tacit government support and a degree of admiration from some sectors of society. By the 1990s however, public attitudes had hardened, and many viewed mercenaries as murderous criminals rather than swashbuckling romantics. One of the last of the old-style mercenary outfits, the South African company Executive Outcomes, mounted military operations in support of the Angolan and Sierra Leonean governments including the use of attack helicopters. This attracted widespread condemnation, especially from the post-apartheid South African government which, in 1998, passed legislation against mercenary activities and the company was wound up.

One of Executive Outcome's last jobs was to supply weapons and ex-special forces troops to the British firm Sandline International, to support the Papua New Guinea government against a rebel movement. The deal caused outrage and Sandline's chief executive, Tim Spicer, a well-connected former Scots Guards officer with driving ambition, was arrested but later released. The episode hadn't blunted his appetite for the murky world of the arms trade. The same year he was contracted to supply weapons to the Government of Sierra Leone. This too ended badly with claims that Sandline had breached an arms embargo. Spicer countered that his activities were not illegal and were in the full knowledge of the British government. Sandline was wound up, but, as we'll see, Spicer would go on to make a fortune in Iraq.

PMC Start-ups

The other side of the guns-for-hire business is more about discretion than aggression. Its origins lie in the North African deserts in WWII where the lofty, 6-foot 6-inch, Scots Guards officer, David Sterling formed the Long-Range Desert Patrol Group. That morphed into the Special Air Service (SAS) creating an elite branch of the military known as Special Forces, or SF for short. Sterling was eventually captured in Tunisia and spent the rest of the war incarcerated in the infamous Colditz Castle. After the war he left the army and moved to Africa before returning to England in the 1950s where he was involved in TV syndication. The glamour of TV could not match the excitement of soldiering and in 1967 he founded Watchguard International, a PMC that provided body-guarding services, training, and security advice in the Middle East and Africa.

Watchguard was the first of many PMCs run from premises in London by former SAS officers. Others included Kilo Alpha Services run by Ian Crooke; Keenie Meanie Services and Saladin, both run by Andrew Nightingale; Control Risks run by David Walker; and DSL run by Alistair Morrison.

There were a handful of foreign PMCs, but the business was, and remains, dominated by London. There are many reasons for this. On the supply side, there was the SAS which, by the 1960s, had wide operational experience having seen action in Borneo, Malaya, Cyprus, Kenya, and elsewhere. Many SAS officers were seconded to the Sultan of Oman's forces and led local troops against the rebels in the Dhofar region. What set them apart from other soldiers, aside from their fighting skills, was their ability to operate independently and to build relationships with governments, businesses and with indigenous people, their quiet determination, and their low-profile approach.

On the demand side foreign governments wanted specialist training, protection, and planning capabilities while multinational companies wanted to secure their people and their infrastructure. This customer base built steadily from the 1960s as the post-colonial period created instability in Africa and in the Middle East with new leaders and their nascent government structures attempting to reconcile colonial borders, internal rivalry, and Cold War power struggles. In the 1970s oil and gas exploration in volatile countries including Libya, Nigeria, Angola, Colombia, Saudi Arabia, and Iraq, needed security support. When in 1973 the price of oil leapt more remote fields became economically viable and the industry needed protection in yet more challenging locations. In the 1980s and 90s world markets became increasingly globalised and multinationals needed advice on how to secure their investments and their staff. And, as we'll see, in the 2000s, international terrorism, as well as the Afghan and Iraqi wars, turbo-charged the demand for PMC security services.

The supply of, and the demand for, international security services came together in London. This was home to government, embassies, oil and gas companies, the global financial and insurance business, law offices, multinationals, and US firms making their first steps onto the international scene. No other city had this concentration of global connections and money. In the PMC world, trust is critical, and personal connections are key.

SAS officers, ambassadors, senior executives, bankers, lawyers, and business leaders are largely drawn from the elite educated at the most prestigious public schools such as Eton, Harrow, Charterhouse, Rugby, and Marlborough. These schools were established to cater for sons of the ruling classes, military officers, and colonial administrators, who wanted a stable education for their children while they were away running the empire. They had a strong emphasis on crown and country, on sport and discipline, on cold showers and the birch. They produced men with an easy confidence, bags of charm, and impeccable manners. They succeeded in being both impressive and irritatingly smooth.

The brightest tended to go on to Oxford or Cambridge then to The City for a career in banking. The sportier went straight to Sandhurst and took commissions in a smart, socially exclusive, Guards regiment, with the cream trying for SAS selection. Well-fed bankers were envious of dashing soldiers, reflecting on Samuel Johnson's famous words, "Every man thinks meanly of himself for not having been at sea or having been a soldier." They wanted some of the glamour and swagger of their old school chums. The SAS officers envied the banker's money and their up-market lifestyles. The two groups formed partnerships of mutual interest, bound together by old school ties and sealing the deal over dinner in the Special Forces Club, the one place in London that bankers couldn't buy their way into.

As a business model, providing security services in far-away lands tends to be feast or famine. Paying for a London office and its staff who drew up contracts, booked travel, paid wages, recruited and administered people, was a constant challenge. Profits for fledgling PMCs could be lucrative, but often they were of short duration before the flow of cash was abruptly turned off. An ideal way of smoothing out the cash flow was to join forces with the insurance business and offer kidnap and ransom insurance.

K & R, as it is known, gained a market in the 1970s following high-profile abductions by ETA in Spain, the Red Army Faction in Germany, the Red Brigades in Italy, and by criminal gangs in Latin America. Protecting people, or negotiating with kidnappers and paying ransoms, demanded specialist skills and piles of cash. K & R insurance provided instant access to both. And PMCs were paid a retainer that solved their cash flow problems.

Organisations would also get a rebate on their premium to spend on security assessments carried out by the same PMC that included advice on how to reduce the risks of kidnapping, and therefore of premiums being paid by the insurance company. This gave the PMC an opportunity to provide more security services. So, for an annual fee, if one of their executives were kidnapped, they would get the services of a PMC staffed by former elite soldiers. It was win-win-win: the risk of kidnap was reduced, there was less prospect of an insurance pay-out, and the PMC received a steady income. The business model was pioneered by David Walker, the former SAS officer who, as mentioned, formed Control Risks in 1975, building a business alliance with the established insurance and travel company, Hogg Robinson.

Stable finances provided a platform for Control Risks to grow an intelligence capability that it sold on a subscription basis to companies who sought analysis of new markets and environments. And once they ventured into them, Control Risks was also on hand to offer specialist advice, risk assessments, executive protection, and security support. As a trusted partner, impeccably discrete, highly professional, with a sustainable business model, it was perfectly placed to support governments during the war on terror.

Boom Years

In chapter 8 we saw how, as the US government sought to reassure a traumatised public, the 9/11 attacks boosted the fortunes of private security companies (PSCs). The attacks also prompted the US government to invade Afghanistan where Al Qaeda was based, and Iraq, where it didn't exist at all. Both wars created a huge demand for PMC services. Operating in Afghanistan and Iraq presented enormous challenges, and for firms capable of overcoming them there were clearly enormous rewards. For a PMC to be competitive, and to be considered a major player in the increasingly global security market, it was essential to get in on the action in these new wars. There were fortunes and reputations to be made. This was El Dorado for the guns-for-hire business.

By far the biggest client for PMCs was the US Department of Defense. The wars stretched the financial and human resources of even the mighty US military. Secretary of State for Defense Donald Rumsfeld was keen to promote, as he expressed in a 2002 article in Foreign Affairs journal, "a more entrepreneurial approach, one that encourages people to be proactive, not reactive, and to behave less like bureaucrats and more like venture capitalists." Value for money was a big factor. The US Congressional Budget Office estimated that a PMC unit was about 10% cheaper than a comparable military unit in the field.

And once their job was done, unlike a regular military unit, their contract was simply terminated.

Individual PMC gun slingers were paid much more than their military counterparts, but they were not paid whilst on leave, they had no pension contributions, and they could be laid off at short notice. Regular soldiers had a low base pay, but their allowances were reasonable, and they retained their jobs when their units were sent back home along with subsidised housing, educational grants, a pension plan, and long-term career prospects.

PMC contractors ready to roll.

Another major factor was headcount. What were supposed to be swift operations to change regimes and instal democratic governments, turned into quagmires of never-ending wars. George Bush's approval ratings hit over 90% immediately after 9/11 but steadily ebbed away to 25% by 2008, as more Americans came home in coffins draped with the stars and bars.

Key performance indicators for the wars were head count and body count. Only serving military were counted, whereas contractor numbers were invisible. This allowed the war effort to grow without the full scale of commitments, and casualties, being immediately apparent.

As well as security, most contractors were involved in base support which included catering, laundry, cleaning, translation, maintenance, logistics, construction, and transport; all areas where civilians were cheaper than soldiers, or where specific skills were needed to help the military to operate. By 2009 in Afghanistan there were 100,000 US troops, supported by 115,000 contractors of which 15,000 were PMC staff. In Iraq that same year there were 115,000 US troops supported by 112,000 contractors of which 12,000 were from PMCs.

There was nothing new about the use of contractors. The US military had used them since the Revolutionary War. In WWII about 10% of the US military's operational capacity was provided by contractors. What was new in Afghanistan and Iraq was the proportion of contractors to soldiers and the number of them that were armed.

The US government was not alone in engaging contractors. Other governments used PMCs to protect their embassies and diplomatic staff. Prior to the US-led invasions, foreigners in Baghdad had few security concerns. Afterwards they became targets for bombings, assassinations, and kidnaps, which demanded sophisticated security measures. Humanitarian organisations, normally averse to hired guns, recognised that if they wanted to help people, they would have to swallow their distaste. They may have had good intentions, but insurgents viewed humanitarians as part of the US-led invasion. Then, from 2008, Iraq's oil fields opened to foreign investment, and international companies also had complex security challenges. In short, any foreigner and foreign organisation was a target, and every foreigner needed protection. The military was too busy fighting to help them, so the gap was filled by PMCs.

Bubble and Bust

PMCs in Iraq and Afghanistan did not provide an offensive fighting capability. Most of their work was guarding military camps and embassies, providing escorts for supply convoys, or close protection for high-ranking officers and officials. A smaller, but significant proportion, maintained situational awareness for logistical and reconstruction efforts. This involved monitoring local events, understanding the security situation, and conducting risk assessments. It was essentially an intelligence and analysis capability. Alongside this were a variety of liaison services, sharing information, and coordinating efforts between the military and civilian organisations.

There were three broad types of contractors providing these services: established defence contractors, who expanded their remit to include armed security services; PMCs for whom armed security was their traditional bread and butter; and other private security companies (PSCs) keen to break into the lucrative new war-zone market.

The major defence contractor was Dyncorp, which had supported the US Government with aviation and logistics services since the 1940s. It was well placed to expand into the provision of security services, including close protection teams for the Afghan President Hamid Karzai.

PMCs, whose business had hitherto occupied a modest niche, were also well placed to make the most of the new opportunities. In 2005, PMC contracts, according to a report by the US Special Inspector General for Iraq Reconstruction, were worth about $6 billion. *The Economist* estimated that British PMC revenues alone increased from $320m before the war to more than $1.6 billion by early 2004. That was just the start. By 2009 the PMC market in Afghanistan and Iraq was worth in the region of $20 billion a year.

The demand was so great that even start-ups were soon able to win massive contracts. The US PMC Triple Canopy, for example, formed in 2003 by former US special forces officers, had by 2005 picked up work from the US government in Iraq, including a multi-year contract worth $1 billion to guard the enormous American embassy in Baghdad.

More surprisingly British start-up, Aegis, won a contract worth over $430 million deep inside the US defence establishment, to provide intelligence and analysis capability. It was headed by the same Tim Spicer who had attracted so much publicity in Papua New Guinea and Sierra Leone with Sandline.

Control Risks was also swift to capitalise on the opportunities. By 1998 it had grown steadily with offices in a dozen countries, a couple of hundred staff, and an income of $20 million a year. Contracts with the US, UK and other governments in Iraq helped them expand rapidly. By 2015 they had offices in 36 countries, 2,500 staff and revenues of $300 m.

Before long there were so many PMCs operating in Iraq that they formed The Private Security Company Association of Baghdad. It was a vehicle for collective liaison with Coalition Forces and with the Iraq authorities and to exchange information on negotiating the local bureaucracy. It had more than

40 members drawn mostly from UK and US companies including ArmorGroup, Hart, Erinys, Kroll, Blackwater, Edinburgh International, and Pilgrims.

The major PSC players, the Canadian Gardaworld and the British G4S, cast envious looks upon this lucrative and high-status part of the security business, and decided to get in on the act. The problem was that they had little form in war zones. Providing guards for supermarkets and hotels in Toronto and Birmingham did not equip them to run diplomatic security in hot conflict environments. Their solution was to buy established PMCs with existing contracts.

In 2006 Gardaworld bought US PMCs Kroll and Vance International which had contracts in Iraq. Keen for more business, in 2009 they entered a low bid for a big contract to guard the British Embassy in Baghdad in a deal worth more than £100 million over three years. The next purchase was Aegis in 2015 for £130 million, which provided a handsome pension pot for Tim Spicer.

Following the Gardaworld lead, in 2008 G4S bought Armor Group, a US company with major US government contracts in Afghanistan and Iraq. Armor Group's origins, as the name suggests, lay in the supply of armoured vehicles and had expanded in 1997 by buying the UK PMC, Defence Systems Limited (DSL) for £27 million.

DSL was another of the early PMCs, established in the 1990s by suave, charming, former Scots Guards and SAS officer Alistair Morrison. Morrison's career highlights included fighting for the Sultan of Oman and assisting the German anti-terrorist unit GSG9 with the release of passengers on board a hijacked plane in Mogadishu. He went on to found DSL along with Richard Bethell, a cigar-chomping aristocrat with a lion's mane of blonde hair, also ex-Scots Guards and SAS. They were both serial PMC entrepreneurs. Morrison went on to establish Erinys which won a major contract to protect Iraq's oil pipelines. Bethell founded the Chelsea Group of security and logistic companies, worth an estimated £500 million by 2020.

After G4S's bought Armor Group they were able to use their corporate weight to bid competitively for major government contracts including the US and British embassies in Kabul.

Private equity was also keen to join the action, setting up Constellis, an umbrella company that from 2011 onwards expanded by purchasing a range of established PMCs including Triple Canopy, Olive Group, Edinburgh, and the infamous Blackwater (rebranded as Academi). Their slogan was, "we go where others won't, and do what others can't."

Gardaworld, G4S, and Constellis were able to use their corporate might, slick administration, and economies of scale to compete for contracts, forming tough competition for the smaller PMCs. But the big players had entered the market at its peak.

By 2011 the US taxpayer was getting tired of the cost of the never-ending wars. Both the Afghan and Iraqi governments became increasingly concerned about the activities of PMCs on their soil. They also wanted a slice of the PMC market, and they were keen to replace foreigners with locals.

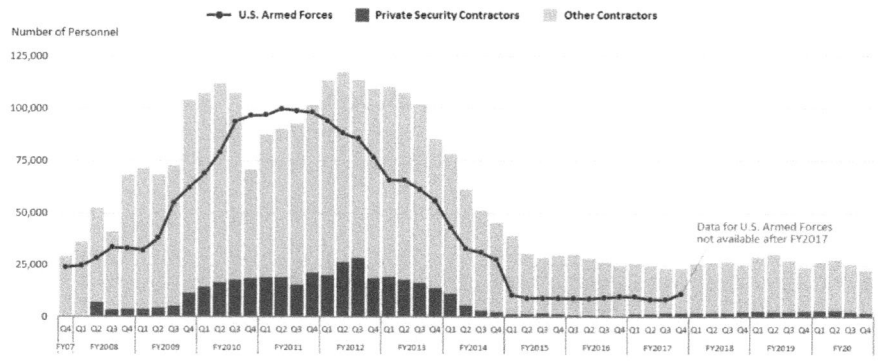

The rise and fall of PSCs in Afghanistan.
(CRS: Department of Defense Contractor and Troop Levels in Afghanistan and Iraq: 2007–2020. 22 February 2021)

The number of foreign PMC staff in Iraq peaked at 15,000 in 2009, dropping to 2,000 in 2012, while in Afghanistan it fell from 28,000 in 2012, to 2,000 in 2015. So, within a couple of years of big PSCs buying their way into the PMC arena, the market shrank precipitously. On top of this, the Iraqi and Afghan governments made life difficult for foreign PMCs by slowing down the issuing of visas, operating and weapons licences, and the importation of equipment. Their profit margins were squeezed, but as entering the market had boosted their credibility, they couldn't abandon their high-profile government contracts without losing it, along with a lot of cash.

The consequences of the PMC bubble bursting were felt at ground level. Wages were the major part of the PMC bill, so to balance the books pay rates dropped from $1,000 a day in 2005 to $300 a day in 2015. To further reduce costs, many western staff were replaced with what were known as TCNs (third-country nationals). TCNs were drawn from lower-income countries with effective military forces, where English was widely spoken. This opened up opportunities for ex-soldiers from Nepal, Kenya, Uganda, and Peru, who would work for just $80 a day. Local nationals also started to be hired and engaged in less sensitive tasks, such as guarding outer perimeters of compounds, and were content to earn $40 a day.

TCNs from Nepal on guard in Kabul, better value for money than westerners.

By 2015 westerners were still employed in management positions, and for protecting western diplomats, but the days of them manning guard posts, and riding shotgun on convoys, were over. The burst bubble left many sitting back home looking for new lines of work and flicking through old copies of Soldier of Fortune magazine.

Blackwater Blackname

Most PMCs did a good job in extraordinarily testing environments, planning carefully, operating cautiously, and protecting those in their charge, sometimes displaying great courage. Others responded to the unrelenting pressure of being in an active war zone by being overly aggressive. Some left indelible stains on the reputation of the PMC sector.

Concerns were already being expressed well before an infamous incident involving the US PMC Blackwater. "They shoot people, and someone else has to deal with the aftermath. It happens all over the place," Marine Brigadier General Karl R. Horst told The Washington Post in 2005. "These guys run loose in this country and do stupid stuff. There's no authority over them, so you can't come down on them hard when they escalate force".

Blackwater was a relative newcomer to the PMC sector. It had been set up as a law enforcement training company in 1998 by Erik Prince, the well-connected son of one of the richest men in Michigan. Prince served in a US Navy special forces SEALS unit for five years before starting the company. Blackwater soon capitalised on opportunities in Iraq, winning a prestigious contract to protect Paul Bremmer, the US Presidential Envoy and by 2010 it had earned more than $1.6 billion in US government contracts.

In November 2007, a Blackwater team opened fire in Nisour Square in Baghdad killing 17 civilians. There were immediate and long-lasting consequences. The outraged Iraqi government revoked Blackwater's licence to operate, and Prince was forced to step down from the company. Four of the Blackwater team were tried for murder in the US where the court heard that, "None of the victims was an insurgent, or posed any threat to the [Blackwater] convoy." They were sentenced to 30 years in prison but were released by President Trump in January 2021.

The Nisour Square incident cast a dark and enduring shadow over the whole sector. There was widespread international condemnation, a determination to prevent a repetition, and demands for the PMCs to be banned or at least regulated. Other PMCs were appalled and pointed to the trigger-happy culture that they believed Blackwater had allowed to fester.

There was no excuse for what happened in Nisour Square, but there is a perspective that helps us to understand the context. One of the reasons that PMCs were engaged in the first place was that if they were killed, they were not included in the official casualty statistics. A research paper by Brown University calculated that in Iraq and Afghanistan, between 2001 and 2019, 7,402 civilians contracted to the US Department of Defense were killed, compared with 6,870 US military deaths. It's not clear how many PMC contractors are amongst these, but it could be as much as a third—so well over 2,000, an average of 2 or 3 a week. Most of these were killed by roadside bombs, in shootings, or in vehicle crashes.

In one of the worst incidents in Iraq, four contractors working for Blackwater were attacked by a mob in Fallujah whilst escorting a food convoy. They were beaten to death and strung up on a bridge over the Euphrates. It was clearly a very dangerous environment that often demanded split-second, life or death judgements.

Unlike the military whose casualties are publicly honoured, PMC casualties are only privately mourned. In Ken Loach's film, *Route Irish*, about a PMC in Iraq, a company executive says, at the funeral of Frankie, a hired gun who was shot dead, "The skills that he learnt were used to protect engineers bringing water and electricity to the long-suffering people of Baghdad, to protect doctors, surgeons, experts in child nutrition, journalists, and electoral advisors. He was a protector, a nation builder, a force for good, and I'm ashamed to have to say this, but there are no memorials in this country to men like Frankie, no Union Jacks, no sympathetic words from politicians, in fact precious little respect at all. In my book these men are the unsung heroes of our time, patriots, soldiers for peace." It is not a perspective that is often aired.

Gun Control

Over the years, many attempts have been made to regulate mercenaries. A 1977 addition to the Geneva Conventions defined a mercenary as a person who is recruited to fight in an armed conflict, takes part in hostilities, is motivated by private gain, and is not a national nor a member of the armed forces of states party to the conflict. It stated that mercenaries do not have the right to be treated as prisoners of war, but their use was not actually prohibited under the Conventions.

Following the activities of characters such as Hoare and Denard, the Organisation of African Unity adopted a Convention for the Elimination of Mercenarism in Africa, which came into force in April 1985 and was adopted by the UN in 1989. But this, too, only outlaws mercenaries who use armed violence to oppose "a process of self-determination, stability or the territorial integrity of another State." This means that a legitimate government can still employ mercenaries.

However, most of the PMCs that provided services in Iraq and Afghanistan fell outside the scope of the OAU, UN and Geneva Conventions, which focus on people recruited to fight, rather than those recruited to protect.

Following the Nisour Square incident, there were urgent calls to regulate all types of guns for hire. The result was four separate initiatives, two state-driven and two industry-driven. On the state side there was the 2008 Swiss-sponsored intergovernmental process known as the Montreux Document which makes it clear that states have an obligation to uphold international law. It makes PMCs accountable to the state where they operate, the state that buys their services, and the state where the PMC is incorporated. It also makes it clear that corporate directors have "command responsibility", so they are accountable for the actions of their staff. The other state initiative is through the UN's Working Group on Mercenaries established in 2005, which monitors mercenary-related activities, advocates for accountability, and focuses the attention of member states.

The state initiatives did not provide guidance on PMC operations apart from a broad obligation to uphold international humanitarian law. However, the industry needed to address the issue of cowboy operators who gave everyone a bad name. The result was two industry initiatives.

The first was a set of guidelines issued jointly by the American National Standards Institute (ANSI) and the US-based security industry association ASIS. The idea was that PMCs could use them as a template for good practice. The second was by a new PMC voluntary body known as the International Code of Conduct Association (ICoCA). ICoCA laid out a set of standards including on recruiting, vetting, training, the storage of weapons and rules of engagement. PMCs joining the association agreed to adhere to the standards and to be subject to a monitoring and evaluation regime. The aim was for users of PMC services to engage only companies which subscribed to the code of conduct. So, whilst it was self-policing, it had a close interest in promoting standards across the whole sector.

Carrying lethal force in a foreign country is always going to be contentious, and past mercenary activity and the Blackwater scandal continue to colour perceptions of PMCs. The issues of accountability, regulation, and legal status remain slippery, but the state and the industry initiatives appear to have raised standards across the PMC business and helped to shine a light on some of its murkier sides.

In the past 50 years, the guns-for-hire business has transformed from wild-eyed adventurers making a quick buck, to smooth corporate executives making large fortunes. Along the way, it's been absorbed as a prestige part of the wider private security business. It continues to be controversial, but standards and levels of accountability have risen significantly. There remains steady business in guarding embassies, humanitarian operations, oil and gas facilities, high-profile individuals and, more recently, endangered wildlife. But the boom years are over. It is easy to imagine the knights of the PMC round table dining at the Special Forces Club, reflecting on the run of fortune that they had, raising a glass to themselves, and saying, "we dared, and we won."

11

FLAK JACKETS NOT DINNER JACKETS

A convoy of three 4x4 vehicles drives swiftly down a potholed road. In the middle vehicle, an armed guard sits up front next to the driver, and two nervous-looking diplomats are in the back. They are all wearing body armour and helmets which makes them uncomfortable and restricts their movements. They scan the road ahead. They know something is about to happen.

BANG! Bang, bang, bang, bang, bang, bang. A large explosion is followed by automatic gunfire. Pulse rates leap as adrenaline flushes through the vehicles' occupants. The convoy abruptly stops and there is a blur of activity. The lead vehicle is hit and enveloped in smoke. In the middle vehicle, the driver shouts "contact wait out!" into a radio. The guard barks "get down!" at the diplomats who breathlessly press themselves low on the back seats.

The guard cocks his weapon, opens his door and uses it for cover as he returns fire in the direction of the ambushers. The rear vehicle pulls alongside the stricken vehicle to rescue their colleagues. Once they are safe inside, the guard in the diplomats' vehicle stops firing and leaps back inside, yelling "reverse!" The two remaining vehicles speed backwards for 100 metres. Then they pull a violent J turn and accelerate back the way they came.

This is not some war-torn country, it's a training ground in Virginia, USA. The diplomats are on a security training course before deploying to an embassy in a war zone. The training is realistic to sensitise them to situations that they might encounter on their postings.

Diplomats are sent abroad to represent their countries, report on the political situation, understand their hosts, maintain positive relations, influence decisions, and get to know the locals. Most are posted to peaceful parts of the world where they can enjoy a privileged life with fascinating work, elegant accommodation, and unique access to the highest echelons of government. But a fair proportion work in war zones, swapping dinner jackets for body armour, where they are more likely to be served a Molotov Cocktail than a Gin and Tonic. Their families stay in their home country whilst they live in windowless bunkers, on embassy compounds. Protecting diplomats in what are known as *kinetic environments* presents an enormous challenge.

This chapter is largely about US diplomatic security, during four key periods. The first, is the Cold War, where there was a focus on countering espionage. The second was Beirut in the 1980s where a series of vehicle bomb attacks against embassies demanded a change in diplomatic security priorities. The third period followed the 1998 Al Qaeda bomb attacks against US embassies in Nairobi and Dar es Salam that resulted in massive investment in embassy security. The fourth period opened with the invasion of Afghanistan and Iraq which resulted in the militarisation of diplomatic security.

Cold War, Hot Spies

Until the Cold War diplomats rarely needed to consider security. Attacks on embassies were rare, and, while espionage had occurred since ancient times, it was not regarded as a major threat. The drawing of the Iron Curtain, and the subsequent superpower rivalry, made military, technical, and economic information, a valuable commodity, and a shadowy game commenced that was immortalised in a whole genre of spy novels and movies.

Since 1948 US Marines have been deployed to US embassies to guard access to the chancery: the heart of the embassy. It's where the Ambassador has his office, where political officers make their assessments, and where classified information is stored and transmitted. Immaculately dressed, the Marines greet staff with inscrutable politeness. They are rarely seen outside the embassy apart from once a year to host a legendary event on the diplomatic calendar around the world: the Marine Corps ball. After the customary flag and drill ceremonies, the dining and drinking begins and by the end of the evening even the most rigid Marine loosens his collar and pulls shapes on the dancefloor.

Once past the Marines, you have to negotiate a series of steel doors each with a combination lock that has a large dial with numbers etched along its rim, like those used on old-fashioned bank vaults. Chancery staff try to avoid being the first to arrive in the morning to dodge having to open up, as it's such an anxiety-inducing start to the day. You must remember the combination (writing it down is forbidden) which is five pairs of digits. You carefully turn the dial five times to the right and line up the first pair of numbers with a mark on the door, then four turns to the left to line up the second pair. Then three turns to the right, two to the left, and finally one to the right. The concentration makes your forehead bead with sweat. You hold your breath as you enter the last pair of numbers hoping for a satisfying click that allows the heavy door to glide open. But it is an unforgiving mechanism. Any slight misalignment and you hear nothing but your own expletives. But that's the point of these locks; they are designed to be difficult to open, to slow down an intruder who might have obtained the combination.

From the late 1940s, embassies also had security officers whose main task was changing combinations on safes, locking windows, checking seals on diplomatic bags, managing encrypted communications equipment, and patrolling offices for unsecured classified information. They had a status similar to traffic wardens, feared but not respected.

Rather like getting points for motoring violations on a driving licence, security officers had the power to give errant diplomats points for any breach of security: two for leaving confidential papers on a desk overnight and three for not securing a window. Accumulate nine points and you'd be sent home, shame-faced, leaving behind your privileged position, generous allowances, and professional reputation.

Security officers may not have been popular, but they were dealing with an active threat. In 1945 the USSR presented the US ambassador to Moscow with a token of friendship: a beautiful wooden carving of the Great Seal of the United States. The ambassador hung it in his study until 1952, when it was found to contain a listening device.

More eavesdropping attempts were to follow. When a new US embassy was built in Moscow in the 1970s, dozens of listening devices were discovered in the main columns, which had been built by a local firm. To prevent a repetition, when a new British Embassy was constructed in Moscow in the 1990s, all the materials were shipped from the UK in sealed containers, and it was built by vetted British workers.

A key vulnerability at embassies was its own diplomats who could be blackmailed into revealing classified information. In the 1950s Senator

Joseph McCarthy fuelled paranoia about the red threat, claiming that the US government was infiltrated by communist sympathisers. As a result, an Executive Order was passed requiring background checks on all federal employees. As well as being a commie, grounds for dismissal included "infamous, dishonest, immoral, or notoriously disgraceful conduct, habitual use of intoxicants to excess, drug addiction, or sexual perversion."

The *habitual use of intoxicants* clause was loosely applied. Diplomats are expected to drink for their country. It's sometimes regarded with envy, but they are often obliged to attend ten or more official receptions a week, gripping and grinning, glass in hand. It takes its toll. A colleague once summarised the hazardous chain reaction of diplomatic life as, "Protocol, alcohol, cholesterol—hospital."

Sexual perversion included homosexuality. The concern was that homosexuals were more susceptible than their straight counterparts to blackmail and were therefore a security risk. Hundreds of US Foreign Service personnel lost their jobs in what became known as "the purge of the perverts."

In 1967, the year that homosexuality was legalised in the UK, the Foreign Office took the view that, "With time, public opinion here, and the laws abroad, may evolve to such an extent that we too can take a more relaxed attitude towards homosexuality, but this time has certainly not yet come." It was not until 1991 that attitudes had evolved sufficiently for homosexuals to be allowed to become diplomats.

The Russians routinely tempted diplomats into *immoral, or notoriously disgraceful conduct*. In 1968, the British Ambassador to Moscow, Sir Geoffrey Harrison, was given a one-way ticket back to London after being honey-trapped by a female KGB agent. In 1987 two Marines at the US embassy in Moscow were also compromised after local romantic relationships led to requests to plant listening devices in the embassy. They too were soon on a plane home.

Bugs and honey pots remained hazards of diplomatic life, but from the early 1980s there was a new requirement—protecting embassies from terrorist car bombs.

Car Bombs

You can pack a lot of explosives in a car. A hundred kilos of it will fit in the boot. A charge of that size produces a blast wave of supersonic gas that can flatten a building 30 metres away and punch out windows 300 metres away. Its effect is almost instantaneous. A rifle bullet flies at 800 metres per second, a blast wave is almost ten times quicker, detonating at 7,000 metres per second. The car itself is shredded into fragments of metal and glass that lacerate any flesh in its path.

A vehicle carrying a bomb is indistinguishable from any other on the road, and it can be driven to any destination. A car bomb is cheap to make, costing just a few thousand dollars, that's why they are known as the poor man's artillery.

Beirut has the unfortunate distinction of being the first capital to witness the use of suicide car bombs to attack embassies. During the Lebanese civil war there were a series of appalling bombings, starting with the Iraqi embassy in 1981 which killed the ambassador and 60 others. In 1982 the French embassy in Beirut was bombed killing 14. In 1983, terrorists used a van carrying 900kg of explosives to bomb the US Embassy. The seven-storey building collapsed killing 64 people. A few months later simultaneous truck bombings killed 241 US Marines and 58 French troops in their respective barracks. And in 1984 the US Embassy annexe was bombed, killing 24.

These attacks where a horrible demonstration of both the effectiveness of vehicle bombs and the vulnerability of embassies and forced a change of security emphasis from protecting information from spies, to protecting diplomats from terrorists. A US Congressional Commission was formed soon after the Beirut bombings under Admiral Bobby Ray Inman. It was tasked to consider how to respond to what he called, "the beginnings of calculated terror campaigns, psychological conflict waged by a nation or sub-group against our nation, with an ever-broadening range of targets, weapons, and tactics."

The 1961 Vienna Convention on Diplomatic relations obliged host countries to ensure the security of their diplomatic guests and to protect embassies "from intrusion or damage." But preventing a determined group from attacking an embassy was beyond the capability of many hosts. It was clear that if bomb attacks could not be stopped, the emphasis had to be on protecting against their effects.

This was no easy task. The most important factor in embassy protection was the distance between a bomb and a building. The greater the distance, or *stand-off*, to give it its technical name, from a bomb, the greater the chances of a building surviving. Few embassies had that sort of real estate surrounding their city centre locations. Whatever the cost, the Inman was determined to prevent the loss of another embassy and prescribed a series of far-reaching security measure including a minimum stand-off of 100 feet (30 meters), ram-proof perimeter walls and tightly controlled vehicle access.

One of the main causes of death and injury in an explosion is flying glass. A blast wave shreds windows and forms a lethal blizzard of glass. Inman therefore, specified that window size had to be minimised and all glass had to be toughened to prevent it from shattering. A major concern was of building collapse as had happened in Beirut. Traditional masonry structures were especially vulnerable. A brick wall acts like a sail, billowing when hit by a blast wave and crumpling in a smoking heap. Inman therefore stressed the need for concrete construction that would withstand a blast and minimise casualties. The age of the bunkerisation of embassies had begun.

As well as new standards for embassy buildings, Inman recommended an uplift in resources for diplomatic security and the development of significant counter-terrorist capabilities. The diplomatic security staff increased from 500 and a budget of $27 million in 1980, to 1,200 and a budget of $300 million in 1986. The US's became, and remains, by far the largest, and the most professional, of all diplomatic security organisations.

East African Attacks

It might have been due to the improved embassy security, or to the easing of political tensions as the Lebanese civil war petered out, but no other US embassies were bombed in the 13 years following the publication of Inman's report in 1985. As time passed the Beirut bombings were regarded as a phenomenon of their time and place. There was a halcyon period between the end of the Cold War and the start of the War on Terror when the world looked rosier and full of opportunity rather than menace. As a consequence, the upgrading of embassy security lost momentum and Inman's recommendations were only partially implemented.

Optimism turned to shock in 1998 when two US Embassies in East Africa were hit simultaneously by massive truck bombs, killing 212 in Nairobi and 11 in Dar es Salaam. These were the first major attacks by Al Qaeda, the first suicide bombings in Africa, and the first fatal attacks on US embassies since 1984. They were therefore completely unforeseen.

A Commission was set up under another Admiral, William Crowe, to review the attacks. The Commission found that there was a "collective failure" at multiple levels of government, and that insufficient resources had been dedicated to embassy security. There was a new determination to protect embassies, and this time it would be cast in law. The US Government passed the Secure Embassy Construction and Counterterrorism Act (SECCA) in 1999 which rebooted many of Inman's recommendations. It became mandatory for embassies to have 100 feet of stand-off unless the Secretary of State formally waived this requirement, it further increased the number of diplomatic security staff, and it set aside billions of dollars for a secure embassy construction programme.

American embassies were supposed to be symbols of freedom and democracy. They were designed to be attractive and inviting, to show respect to their hosts, and to foster goodwill and admiration. For example, the US embassy in New Delhi, built in 1959 combined Indian and Western architecture. Its graceful lines are mirrored in a reflecting pool similar to that in front of the Taj Mahal.

Following the 1998 bombings, security and speed of construction, rather than style and symbolism, became the priority. Standard designs in three sizes—small, medium, and large—were developed for all embassies and could be built within 24 months. The new US Embassies looked more menacing than inviting, more Sebastien Vauban than Antoni Gaudi.

To achieve 100 feet of stand-off on all sides, an embassy the size of a tennis court needs a plot of land the size of a football pitch. Achieving this in downtown locations is eye bleachingly expensive even for deep-pocketed Uncle Sam. As a result, many embassies were relocated to the suburbs, or to inner-city wastelands. For example, the London embassy was moved from elegant Grosvenor Square to unfashionable Nine Elms south of the Thames. The Beirut embassy moved from the charming historic centre to an austere 43-acre site 10 miles outside the city. These projects each cost in the region of $1 billion.

The key performance metric of any security programme is the absence of incidents. The problem is that you don't know how many incidents have been prevented in the first place. However, there was evidence that the new Inman standard compliant US Consulate in Istanbul had done its job. In 2003 terrorists believed it was so secure that "they don't let birds fly." As a result, they switched targets to the more vulnerable British Consulate killing 16 including the Consul General, Roger Short, in a car bomb attack.

In some ways, this displacement of target was a quiet triumph of the US approach, but it was a major conundrum for other allies, especially its closest: the UK. As a boots on the ground supporter of US foreign policy, including the war on terror, the UK was in Al Qaeda's crosshairs. The UK had an extensive network of embassies, most of them, as was the case in Istanbul, in old buildings in the historic centres of capitals: good at projecting fading British grandeur but bad at protecting diplomats from car bombs.

The UK responded to the Istanbul bombing by issuing a set of security standards similar to those of the Inman commission. They too emphasised the importance of stand-off, perimeter protection, strengthened windows and strong buildings that could resist collapse. The problem was that the physics of explosive attack were the universal, while the UK's budget for its diplomatic service was less than 5% of the US's. It could not afford to buy slabs of real estate and construct fortresses. If horizontal stand-off could not be achieved, the solution was to go for vertical stand-off in strongly constructed buildings. In Madrid, for example, the British Embassy moved from its elegant downtown building, to the 40th floor of a new tower block on the edge of the city. The building was shared with commercial businesses, firms of accountants, lawyers, and financiers, but at least it was protected from car bombs. And it had a great view.

Expeditionary Diplomacy

The next major step in diplomatic security followed the invasions of Afghanistan and Iraq. Having toppled the regimes, significant diplomatic (and ultimately fruitless) effort was required to stabilise both countries. The nature of the security challenge became apparent early on when, in 2003, terrorists drove a truck bomb into the UN's Baghdad HQ in the Canal Hotel. It killed 21 including the UN's Special Representative, the charming Sergio Vieira de Mello, who had been tipped as a candidate of Secretary General.

If the UN's blue flag offered no protection, the US's stars and bars was a sure target. And as well as vehicle bombs, diplomats risked attack by ground forces, rocket fire, roadside bombs, kidnappers, and violent mobs. Keeping them safe demanded an extraordinary range of security measures, which meant keeping Iraqi people at a distance, and spending a fortune.

Because host government control outside Kabul and Baghdad was so tenuous, embassies bucked the new trend for out-of-city sites. They might be able to identify large plots to provide stand-off to build a fortified embassy, but construction took years, and diplomats would have been vulnerable to insurgents when they ventured outside. As a result, embassies were located squat in the heart of government areas within what was known as the *green zone*: a secure enclave ringed by high concrete walls and barbed wire. The green zone was protected by host nation troops armed with machine guns and tanks, and subject to a curfew between dusk and dawn. Access was restricted to diplomats and to vetted locals who worked inside the zone, and then only after being searched.

A *green zone* may sound like an oasis with shady palms and lush grass. In reality, it is a giant concrete labyrinth where people lurk inside fortified compounds. Within the green zone, all you see are rough concrete walls, five metres high, taller than a double-decker bus. The walls are themselves lined by huge concrete blocks to prevent vehicles from getting too close. You can't see what is behind the walls because the buildings are low and are set back the statutory 100 feet. The concrete is softened only by accumulations of sand and grit in the angles between the walls and the ground. It is a security dystopia that would more accurately be called the grey zone.

Entry to an embassy is via a road lined with yet more concrete; low anti-ram barriers that funnel vehicles into a narrow channel. The barriers are low so that the vehicles can be observed by guards with heavy machine guns, ready to open fire should those inside prove hostile. The road has chicanes and speed bumps that slow vehicles to a walking pace, and spring-loaded ground spikes to prevent reversing.

All vehicles are searched in an *air-lock*, a fortified bay bracketed by pairs of heavy steel barriers that look as if they have been hacked off an oil rig. The bay is surrounded by steel fences to stop anyone who tries to escape, and heavy concrete walls to channel the blast should the vehicle explode. The first barrier opens, the vehicle enters the bay then stops before the second barrier. The first barrier then closes, locking in the vehicle while guards check the passengers' identities and search for explosives and weapons. It's not the warmest of welcomes.

Once inside the compound, the embassy buildings hove into view. They look as if they are made of giant Lego blocks covered with sandpaper. The only windows are slender, like arrow slits, and are fitted with bullet-proof glass. All external doors have thick walls two metres in front of them to deflect blast. The gritty facades are dotted with CCTV cameras, floodlights, loudspeakers, and sirens.

Entry to the various parts of the embassy is monitored by armed guards and controlled using pedestrian *air-locks*: like their vehicle counterparts, these *air-locks* have a double-door mechanism designed to stop tailgating. The first door opens, visitors enter a bullet-proof glass chamber, it closes behind them, and a second door then opens to let them into the next area. It feels more like a prison than an embassy.

Not the warmest of welcomes.

Incoming

Attacking the multi-layered defences of an embassy in a green zone is difficult. The easiest way for insurgents to express their emotions is to mount rocket attacks. Not the sort you use to celebrate New Year's Eve, but military high-explosive rockets, often the Chinese designed 107mm, which fly in a high arc and have a range of about 8 km. Unlike artillery shells, which need a long-barrelled gun, rockets are launched from a short tube or a rail, so they are very quick to set up and fire, and they are fiendishly effective.

A 107mm shell weighs 20 kg, so it's like an anvil falling from a great height. Its blast is powerful enough to destroy a house and its shrapnel can kill anyone unprotected within a 50-metre radius. To stop the shells from penetrating a building, a roof needs to be as hard as woodpeckers' lips and it's usually a metre of reinforced concrete topped by layers of sandbags. To hold up the extra weight of the roof, the building needs massive reinforcement which further escalates costs.

Fortunately, the arrival of a rocket is normally signalled 10 to 20 seconds in advance by C-RAM. This is a Counter-Rocket, Artillery, and Mortar detection and warning system, suspended from a blimp that floats high above the green zone. It identifies incoming projectiles as they whoosh overhead and sends alerts across the green zone. On hearing the alarm everyone has a few seconds to dash to one of the many bomb shelters that dot the embassy compounds and hope for the best.

When on a home posting, diplomats live in their own modest homes in commuter suburbs. When posted to peaceful countries they are allocated attractive city centre accommodation. The joke is that an overseas diplomatic posting is where second-class people can enjoy a first-class lifestyle. By contrast, a posting to a war zone means living on the embassy compound in a spartan bunker.

A diplomat's troglodyte dwelling.

Everyone dines in shared canteens where all food is shipped in from the US or Europe. There are recreational facilities including cinemas, games rooms, gyms, and swimming pools, but it is still a work, eat, sleep existence. A diplomat in this environment becomes a chunk (from eating too much), a hunk (from all the gym time), a monk (isolated in their rooms), or a drunk (from propping up the embassy bar).

The upside of this institutional lifestyle is that diplomats are paid extra allowances that can almost double their pay. And many work a routine known as six and two: six weeks at work, followed by two weeks leave. Married diplomats generally head home, while the single head for a beach in Thailand, Jamaica, or Ibiza, where the extra allowances cover the cost of prawns and daiquiris.

As well as diplomatic staff, contractors who provide security, catering and maintenance services live on the compound. Consequently, embassies in hostile environments tend to be very large. The US embassy in Baghdad covers 42 hectares (the same size as the Vatican) and has 4,000 people living within its walls. It is the largest US embassy in the world, and cost $1.5 billion, as much as the Burj Kalifa in Dubai, the tallest building in the world.

The responsibility for security at US embassies falls to the Regional Security Officer (RSO), a key member of the Foreign Service's Diplomatic Security organisation. The title RSO comes from the early days of diplomatic security when a single security officer covered multiple countries. By the 21st century all embassies had their own RSO, many had several deputies, and in Iraq the RSO is supported by around 200 staff.

Normally ex-military or police, RSOs are like sheriffs, no-nonsense types who are selected for their adaptability, robustness, and willingness to say "no sir" to an ambassador. The RSO organises all aspects of security, crisis management, risk mitigation, staff training and contingency planning for the embassy, as well as liaison with, and training for, the host nation security forces who guard the approaches to the embassy.

The RSO also oversees the embassy guard force which is the size of a small army. The US embassy in Baghdad has 5,000 guards, all provided by PMCs (Private Military Companies) and who live within the compound. As well as those peering out of watch towers, patrolling the perimeter, guarding entrances, and searching visitors, there are control rooms with CCTV operators and operations managers, offices with ID badge administrators, shift managers, planners and intelligence analysts, armourers, technicians, and maintenance staff. It is an immensely complex operation.

Mobile Security

The majority of diplomats posted to Baghdad never leave the compound except to go to the airport, and most never meet a local national during their posting. For those that do venture out, detailed planning and extensive security arrangements are needed. All transport is in armoured vehicles.

Chapter 11: Flak Jackets not Dinner Jackets **167**

These are not tanks, but civilian 4x4 SUVs (often Toyota Landcruisers), that are discretely armoured to protect against bullets and low-velocity fragments. They are difficult to distinguish from non-armoured 4x4s until you weigh them and look at the price tag. A normal Toyota Landcruiser weighs 2.5 tonnes and costs $40,000. The armoured version weighs double that and costs three times as much.

Armoured vehicles provide limited protection against the insurgent's favourite weapon: the roadside bomb. A roadside bomb is typically an artillery shell positioned with its solid steel base pointing towards the target and a detonator in its nose. It is set off by a spotter using a mobile phone which sends an electric pulse to the detonator, or by a sensor like those used in intruder detector systems, that identifies an approaching target.

Armoured diplomatic vehicles ready to dodge bombs and bullets.

A bullet weighs 25 grams and travels at 800 metres per second. The base of an artillery shell weighs four kilos and travels at 7000 metres per second. It's the difference between being hit by a sledgehammer and being hit by a freight train. An artillery shell base plate carves through an armoured vehicle, in one side and out the other, slicing through anything in between. It is devastatingly simple and effective.

The main defence against roadside bombs is intelligence. If you know where bombs are likely to be planted, you can avoid the area. But intelligence is fallible, so armoured vehicles are fitted with an array of electronic countermeasures that interfere with a bomb's initiation mechanism. They can be effective, but they have distinctive aerials, which act like a big sign saying, "high-value target within."

It's cramped inside an armoured vehicle. There are radios, weapons, ammunition, countermeasure control boxes, maps, fire extinguishers, spare parts, first aid kits, bottles of water, and tow ropes. With so much kit piled inside there is just enough room for the driver, a guard riding shotgun up front, and two passengers in the rear. The bullet-proof windows are as thick as a hamburger. Each door weighs 200 kilos—as much as a Sumo wrestler—so when parked on a side hill it's impossible to open the uphill door.

Along with the engine noise, crackling radios, and the rattle of equipment, there is the constant hum of air con. In Iraq, summer temperatures can top 50 degrees and with the sun beating through the windows, it soon becomes bread-bakingly hot unless the air con is set to max.

Everyone wears PPE: personal protective equipment. This consists of 7kgs of close-fitting body armour, and a 2kg helmet. The body armour is rigid like a turtle's shell and restricts your breathing. The helmet makes your neck hurt as you look over the driver's shoulder at the road ahead. Why wear armour inside an armoured vehicle? If there is an incident, and you have to make a swift exit, you need to be geared up. And the helmet prevents your head from bashing against the side windows as the vehicle, with its rigid suspension, pitches into potholes.

By the time you have put on your body armour and walked under the relentless sun to the vehicle, you are gasping for air like a landed fish. You try to maintain dignity as sweat sluices down your back and into the elastic of your underwear. You fear falling over and being pinned helplessly to the ground by the weight of the armour. The close protection team (or CP as they are known) brief you on the journey ahead and go through *actions on* drills. That means actions on breaking down, being attacked, surrounded by demonstrators, or crashing. The bottom line is that you do everything they say without hesitating. The CP then wrestles open the Sumo door, and you twist and fold into the rear seat. The CP watches, evaluating how you might react if there is an incident. Diplomats try to act cool and competent, and the CP tries to restrain his patronising smile.

Solo vehicles cannot venture outside the green zone. If one were to be ambushed, immediate back up is needed, so all travel is in three-vehicle convoys, with the diplomat's vehicle in the middle. The first vehicle looks out for trouble and negotiates checkpoints, and the third vehicle is to provide support should the first two get into trouble. To reduce targeting opportunities the convoy drives furiously fast, skittling away other road users. It may reduce the risk of being targeted, but it wins few friends.

As a young man, I travelled across the Middle East on local buses. Returning as a diplomat I was always self-conscious to have swapped my hippy pants and backpack for a stay-pressed suit and Kevlar armour. Once, when planning a trip from Baghdad to Fallujah to review a project, I told the security manager that I wanted to remain low-key. However, he was proud of his good relations with the Iraqi police who provided escorts outside the green zone. When he told them that his boss was visiting, in a show of generosity, in addition to our three armoured Landcruisers they organised two fearsome Humvees capped by turrets with 50-calibre machine guns that fired bullets as big as your thumb. I ticked with frustration as we sped along the dusty highway to Fallujah.

Before entering the town, we had to pay a courtesy call on the local military commander. He received me with flamboyant grace in his chilly airconditioned office and we chatted over sugary tea. Such was his hospitality that he declared that he would accompany me on my visit.

Our five-vehicle convoy was swollen by the commander's Humvee and a further ten escorting vehicles full of soldiers and their guns. As we stopped in Fallujah, they jumped out and formed a hedgehog around me like an infantry square at Waterloo. How I longed for my carefree backpacking days.

The sad truth is that few diplomats ever encounter a local in the wild. For most, the only meaningful interaction is a short meeting with an official inside a fortified Ministry guarded by a close protection team which nervously scans the corridors for would-be assassins. But the threat is real. In Baghdad in 2007 a British computer expert working in the Ministry of Finance, and his four close protection officers, were abducted. They were all subsequently killed.

It wasn't an isolated example: dozens of foreigners were kidnapped in Iraq and ransomed for money or prisoner exchanges. Any foreigner can fetch a price, but a captured diplomat would carry a premium and generate massive publicity. They are prime targets in need of serious security.

That's me, fourth on the left, having a discrete meeting in Falujah, Iraq.
(Courtesy of Optima)

Red Zone

Green zones were reasonably effective in keeping attackers away from embassies, but they could still be penetrated. In Kabul, insurgents repeatedly breached the green zone's walls, bombing the Indian embassy three times in as many years. The worst attack, in 2010, killed 18 including nine Indian diplomats.

Insurgents struck Kabul's US embassy in 2011. A group of six armed with machine guns, grenades, suicide vests, and fanatical determination, aimed to hold ground for as long as possible. The embassy's layers of defence held strong and no Americans were killed despite the target being struck repeatedly by rocket-propelled grenades. The insurgents, inevitably, were killed, but they achieved their aim of embarrassing Afghan and US security forces, and they dominated world headlines for 24 hours.

A few months later, insurgents used similar tactics to attack the British Council in Kabul and succeeded in breaching the outer defences using two car bombs.

Ten people were killed in the attack, and it took eight hours and a heroic defence by the PMC team before control was regained.

Baghdad's green zone was breached in 2019 by violent protestors who stormed the US embassy. They penetrated the outer defences and set fire to the reception area but did not get as far as the main embassy building. No one was killed in the incident, but it was a close call. Aggressive mobs are one of the hardest security threats to counter. Stopping them often requires the use of force, but lethal force can further escalate the situation. In Baghdad, the combination of good security design and a restrained response by the embassy guard force proved effective.

In 2012 the US consulate in Benghazi, Eastern Libya, was attacked. Unlike in Kabul and Baghdad, there was no green zone. The consulate had been adapted from local buildings rather than specially built. Host forces were unreliable, and US reinforcements were many hours away. It was an isolated outpost, like the Alamo.

Dozens of local insurgents armed with heavy machine guns, mortars, and rocket-propelled grenades overwhelmed the consulate's guard force. The US Ambassador, Chris Stevens, and other colleagues, took refuge in the chancery's safe room but were forced to escape when fire broke out and filled it with smoke. A CIA operator frantically messaged the embassy in Tripoli saying "If you guys don't get here, we're all going to fucking die".

But help was a long way off. As dawn broke and the smoke cleared, the Ambassador was dead along with three US colleagues, seven Libyan guards, and an unknown number of attackers. It was the heaviest loss of US diplomatic staff since the Nairobi bombing of 1998 and the first time that a US Ambassador had been killed since 1976, when Francis Meloy was kidnapped and shot dead by terrorists in Beirut.

The episode demonstrated the risks of conducting diplomacy in loosely controlled bad lands, where armed militias roam free. There were various investigations into the attack. Some concluded that security measures at the Consulate were sub-standard, and a handful of Diplomatic Security officials were forced to resign. Ultimately, the issue was that the Consulate was outgunned by a vastly superior force and the cavalry couldn't get there quickly enough.

Inevitably, the Benghazi attack led to another inquiry which resulted in yet more resources for diplomatic security. Benghazi was a hot political issue, especially for Hilary Clinton who was Secretary of State and a Presidential candidate at the time. Determined to prevent a repetition, funding shot up from $2.7 billion in 2012 per year to $4.5 billion the following year. By 2020 it had further increased to $5.4 billion (as much as Finland's entire defence budget). The number of diplomatic security staff increased to around 3,500 direct hires and a further 41,000 contracted embassy guards (about the size of the Canadian Army).

The US Foreign Service provides security for 24,700 Americans in 170 countries at an average annual cost of more than $200,000 each. The costs vary considerably by country. Places like Iceland and New Zealand need a light security regime. By contrast, in Iraq, it costs around $1 million to protect each of the 1,000 diplomats—a cool $1 billion a year.

There has been a high human cost too. More than 70 US diplomatic security staff and contractors have been killed protecting diplomats since 2000, most of them in Iraq. Theirs is a proud record: aside from Benghazi, only a single US diplomat (Anne Smedinghoff, victim of a suicide bomber in Southern Afghanistan in 2013 while delivering books to a school) has been killed in the two decades of expeditionary diplomacy since 9/11. US Diplomatic Security stepped up to an epic challenge.

However, there are diminishing returns. The more difficult the security environment, the more stringent the protection for diplomats, which results in less interaction with the locals. In war zones diplomats can be kept secure inside fortresses, but their effectiveness is severely blunted. One could ask whether it is better, instead, to have diplomats work remotely from the safety of their home capital. If the pandemic taught us anything it is that a lot of business can continue using Zoom or Skype. It would certainly save a lot of lives and money, and there would be no need to trade in the dinner jacket.

12

ENTER THE LAWYERS

On 8th of August 1840, on the road between Hartford and Holyhead, the wheels on a mail coach collapsed. The driver, a Mr Winterbottom, was thrown from the coach and was injured; "lamed for life," claimed his lawyer. A Mr Wright had been contracted by the Post Office to maintain the coach but, claimed Winterbottom, his work was so shoddy that it left the coach in a "frail, weak, infirm, and dangerous state and condition.… unsafe and unfit for use."

This was a landmark case, one of the earliest attempts to bring a claim of negligence. It was unsuccessful for two reasons. First, there was no direct relationship between Winterbottom and Wright. Both had been hired by the Post Office, but the judges took the view that there was no contractual relationship between them, so they could not sue each other. Second, poor workmanship was so common at the time that, if they allowed Winterbottom to sue, it would open the floodgates for similar claims, and that would not be in the public interest.

Had he been injured in similar circumstances today, it is very likely that Winterbottom would have had a successful claim against Wright. We remember his case because it was a milestone on the road to the modern concept of duty of care, which, by the 21st century had become a major driver of standards in workers' and consumers' rights, and, in our favourite topic, security. There are very few laws that specify standards of security, so duty of care was, and is, a vehicle for defining security measures and responsibilities in a range of circumstances.

Duty of care makes a powerful case. CEOs might play down requests for uplifts from a security manager, but their ears pick up when a corporate counsel explains their exposure to legal, ethical, and financial risks. Security managers and lawyers forged an unlikely romance, the product of which was a more comprehensive approach to preventing both security incidents and associated litigation.

Duty of care has been embraced by the security sector in the past 20 years, yet there was no single issue that drove security into the arms of the lawyers. It was more a gradual flirtation as new cases arrived, each helping to define an organisation's security responsibilities. Duty of care gave security managers a new level of gravitas. No longer did they have to tap the side of their nose and say that bag searches, or travel briefings, or the wearing of body armour were "for security reasons." They had a more noble concept, a magical phrase, a trump card, an unassailable argument that could be delivered with a benign smile: "Duty of Care, Mate." People might push back against an overzealous security manager, but they could be neutralised by this veneer of legal respectability. Not only were they protecting you, they were also protecting the organisation.

Duty of care has its origins in *tort*, which is a civil wrong that causes loss or harm to somebody. A *tort* is not a crime. A crime is when you break the law, and are punished by the state. A *tort* is when you are liable for someone's loss or injury because of your negligence, and you must pay compensation.

The key principle is that *reasonable measures* should be taken *against reasonably foreseeable harm*. For example, if someone slips and injures themselves on a shop floor because a bottle of olive oil has been spilt, the shop owner may have failed in their duty of care by not promptly clearing up the mess. It is not illegal to spill oil on the floor, and it is not illegal to leave it there. But if someone is harmed in this situation, they can seek compensation using the duty of care principle.

To win a case, a claimant has to prove three things. First, that the defendant owed them a duty of care. Second, that the defendant breached that duty of care. And third, that the claimant suffered harm as a direct consequence of the breach. In our example, a shopkeeper has a duty of care for everyone that enters their premises. Failing to clear the spilt oil resulted in a customer slipping and harming themselves. Lawyers would rub their hands at the prospect of such a case. Shop owners would put their head in theirs.

Duty of care is both great and terrible. Great because it is not fixed in tablets of stone. It is a flexible way of testing a case taking into account all its complexities. And it doesn't stand still, it evolves over time, and adapts to circumstances. Terrible because it can only be tested by a court *after* an event, and the judgments are not always consistent. So you can speculate, but you will never know for sure if you are actually fulfilling your duty of care.

In all duty of care cases, money is the key driver: it is the only means of compensating those who suffer loss and punishing those found accountable. Criminals and terrorists could be found guilty of a crime and be punished by law, but they rarely have much money. As a result, claimants look to pin liability on those with deep pockets and who failed in some way to prevent the attack. So, strangely, it is often the protectors, rather than the perpetrators, who end up paying compensation.

Like much of law, duty of care cases rely on precedents. Past cases are beacons for the future as courts seek coherence in their judgments. In this chapter, we'll consider some of the landmark cases that have established precedents. But, for a variety of reasons, often to save time, money, bad publicity, or because the outcome is predictable, many cases don't go to court.

Also, negligence claims are often covered by insurance. Insurance companies are happy to take premiums and reluctant to make pay-outs, so they often seek out-of-court settlements rather than establish a legal precedent that might increase their future liabilities. Fear of bad precedent is a key driver of indemnity insurance.

Many years after Mr Winterbottom's unsuccessful case, duty of care has gained traction and continues to evolve. It remains a nebulous concept and it causes intense debate about what is *reasonable*. This chapter reviews how duty of care unfolded, first in the US, and then in the UK, and considers its profound effect on the world of security.

US: Automobiles and Automatic Guns

The first case in the US that we'll look at has strong echoes of Winterbottom v Wright, but, 70 years later, the outcome was different. In 1909, Donald MacPherson bought a second-hand Buick Runabout from a car dealer in New York. He was injured when a defective wheel collapsed, and he sued the Buick Motor Company for negligence. In their defence, Buick said they had manufactured the vehicle but not the wheel, which they had purchased from another company, and, in any case, the vehicle had been purchased from a dealer, not from themselves.

The court found Buick liable because wheels were an integral part of the vehicle, they should have been checked for defects, and, crucially, Buick knew that their vehicles would be "used by persons other than the purchaser, and.... irrespective of contract, the manufacturer of this thing of danger, is under a duty to make it carefully."

This judgment established the principle of product liability. Winterbottom was ahead of his times. In the 1840s there was little quality control and no concept of responsibility beyond a formal contract. By MacPherson's day, manufacturing standards had improved, and it was normal for consumers to purchase items through agents, distributors, or second-hand, many steps removed from the manufacturer. The old principle of *caveat emptor*—buyer beware—was increasingly challenged in the complexities of the 20th century.

By the 1960s, product liability and consumer safety were widely understood. As with all risks, there was a balance to be struck between costs and benefits, but, as the Ford Motor Company was to discover, not when it came to human life. Ford was concerned that Volkswagen and Toyota were set to dominate the US market for sports saloon cars. Their response was to build a car weighing less than 2,000 pounds and costing less than $2000. Known as the Pinto, it hit the showrooms in 1971 having been rushed through the design and manufacture process in only 25 months, much quicker than the normal 43.

During the development phase, engineers identified a fault. The fuel tank was prone to rupturing and exploding if the car was involved in a rear-end collision. However, the fault was not rectified, the Pinto was released, and it burst into flames so regularly that it became known as "the barbeque that seats four." Not great publicity, but Ford did the maths. Recalling every Pinto (more than 3 million were made) and modifying the fuel tank would have been more expensive than pay-outs to incinerated victims, so they decided to adopt the more cost-efficient compensation strategy.

The decision might have made sense to the accountants, but not to the judges. In 1972, a Pinto burst into flames following a crash in California. The driver, Lily Gray, was killed and her 13-year-old passenger, Richard Grimshaw, was severely burnt. Richard's family sued Ford for negligence. It took nine years for the case to reach trial but, finally, he was awarded $2.8 million in compensation for his injuries and a further $125 million in punitive damages (later reduced to $3.5 million). At the time it was the largest-ever product liability and personal injury case. Ford was savaged for putting profit before lives, and the case demonstrated that organisations must put safety first.

This meant that people should not be put at risk without adequate, what became known as *duty of care measures*, irrespective of their cost. While the general principle was recognised, there were no laws laying out what security standards should be. There are many laws that specify safety standards including for food, electricity, and protective equipment, but nothing on security.

Why is that? It's because safety is about protecting people from things, and things are easy to define and simple to remedy: infections can be prevented by having clean kitchens, electrical shock can be prevented by insulation, broken toes can be prevented by steel toecaps. But security is about protecting people from people and the threat they pose comes in an infinite variety. How do you prevent a stabbing, or a bombing, or a shooting? Prescribing security measures that are effective and universally applicable is extremely difficult. The security industry often uses a principle known as JASPAR (justifiable, affordable, sustainable, practical, achievable, reasonable) when considering security measures. It works well for individual locations facing specific problems, but can't be applied generally without one or more of the aspects being lost. All this means that because security measures can't be cast in law, there is wide scope to use civil tort mechanisms to seek redress after an event.

In previous chapters we saw how the increase in the number of privately owned venues led to an increase in private security, as owners, rather than the police, had responsibility for what happened on their property. Duty of care cases helped to define the standards of security on private property. For example, in Mississippi in 1997 a waitress was attacked and raped in the car park of a motel where she worked. She sued the motel for failing to provide adequate security, or duty of care measures. The motel had hired a private security company that employed only a single guard to look after multiple properties over the 240-acre site, and there was inadequate lighting in the car park. The case was settled before it went to court, for $1.6 million. This example is relatively common in the US. It illustrates how owners are exposed to claims if there is an incident on their property, and that in hindsight many things become reasonably foreseeable.

The World Trade Center (WTC) bombing of 1993 triggered an exploration of the frontiers of duty of care. Al Qaeda terrorists attempted to blow up the North Tower using a fertilizer bomb concealed in a truck. The Tower did not fall as planned, but 6 people were killed and many more were injured. Once the dust had settled, the lawsuits started to fly. In previous eras, an analysis of such a scenario would have led to the straightforward conclusion that the terrorists alone were responsible for the attack.

Across the shifting liability landscape, lawyers went probing for new angles. Terrorists may have planted the bomb, but others had not prevented the attack from happening in the first place. To help understand how the attack happened, the lawsuits considered the layers of responsibility. And amongst the measures to prevent a reoccurrence at the WTC following the attack, was that all vehicles from then on, were X-Rayed in enormous machines before entering the basement.

The WTC was owned and operated by Port Authority of New York and New Jersey which brought a suit against the manufacturer of the fertilizer used in the bomb. The Port Authority claimed that the manufacturers had failed to take steps to render their products safe, pointing to a 1968 patent for a non-detonatable fertilizer. The court dismissed the case saying the "defendants owed no duty to plaintiff and that the WTC bombing was not proximately caused by defendants' actions." Given that fertilizer had routinely been used in terrorist bombings across the world, and that non-detonatable fertilizer was feasible, the judgment might have gone either way.

The relatives of the victims of the bombing filed lawsuits against the Port Authority claiming that security at the WTC was inadequate. These claims were not so easy to dismiss. A New York court's initial judgment was that the Port Authority bore the majority of the liability for the bombing. It held that terrorist attacks were foreseeable and that the Port Authority was obliged to take *reasonable measures* to defend against them.

Terrorist incidents in the US are extraordinarily rare. Before 1993 the last bomb attack of any significance was at LaGuardia airport in 1975 when 11 people were killed (no one claimed responsibility, and no one was charged). Prior to that, you have to go back to the 1927 bombing of a school in Michigan by a disgruntled employee which killed 46, or the 1920 Wall Street bomb delivered by horse which killed 30. In the 20th century, more Americans choked to death on pretzels than were killed in terrorist attacks. So, the courts were pushing the boundaries of what was *reasonably foreseeable*.

The Port Authority took the case to the Court of Appeals and won by the narrowest of margins: four judges against three. It used a defence only available to public institutions and claimed immunity as a governmental body. They got away with it by the skin of their teeth. Had the WTC been privately owned the judgment is likely to have gone against them. The case marked a clear shift of judicial views on how responsibility is allocated in terrorist attacks: the venue owner, rather than the attacker, bears the weight of tort claims.

The litigation arising from 9/11 further pushed the boundaries of what were considered reasonable steps to prevent terrorist attacks. Families of the 9/11 victims sued a variety of entities, including the manufacturers of the hijacked airplanes. The families claimed that, had there been adequate security on the plane's cockpit doors, the attacks would never have taken place. Were they right?

The families argued that hijackings were reasonably foreseeable (worldwide there had been more than 150 in the previous 40 years) and that strengthening cockpit doors was a reasonable and cost-effective measure. Put in those terms, the argument was compelling and in an out-of-court settlements, an average of $1.5 million was paid to each of the 3,000 or so claimants.

It might have been considered *reasonably foreseeable* that a plane would be hijacked, but even so, the odds were pretty long. By 2001 there were in the region of 20 million flights a year, of which an average of two a year had been hijacked in the previous decade. This means that there was a one in 10 million chance of any individual aircraft being hijacked. By contrast, according to a 2009 article in National Geographic "the average American has about a 1 in 5,000 chance of being struck by lightning during a lifetime". Despite hijackings presenting a smaller risk than lightning, they were still judged reasonably foreseeable. But the key metric was that a few thousand dollars spent on modifying the cockpit doors could have prevented a multibillion-dollar catastrophe.

If hijackings were reasonably foreseeable, then surely mass shootings, which occurred almost monthly in the US, were reasonably foreseeable too? In July 2012, in a Cinemark theatre in Aurora, Colorado, during a midnight screening of the latest Batman movie, Dark Night Rises, 25-year-old John Holmes, wearing military clothing and sporting dyed red hair, shot into the audience. Twelve people were killed and 70 were injured.

Relatives of the victims sued Cinemark for failing in their duty of care arguing that the incident was both foreseeable and preventable. Mass shootings in the US were common enough, and the relatives claimed that security measures such as metal detector arches, bag searches, and additional security staff would have stopped Holmes from entering the premises with weapons.

The court, however, ruled that the attack was "completely unpredictable, unforeseeable, unpreventable and unstoppable." It reasoned that to blame Cinemark would have made all businesses liable for events they could not possibly predict. "The inescapable conclusion," the judge said, "was that this was a horrible tragedy."

Cinemark Aurora took no chances and did not screen Joker, another film of the Dark Knight Rises genre, when it was released in 2019. There was concern that its portrayal of a psychopathic loner might inspire violence, especially as there were suggestions that John Holmes, with his red hair, had modelled himself on Batman's nemesis, the Joker. Many other theatres screened it amidst tight security. In the event, there was no violence associated with the movie, but had there been, any resulting litigation would have posed a difficult test for a court. If such a case was successful it may have led to substantial pay-outs, increased insurance, and the prospect of airport-style security at cinemas and other venues across the US.

Another mass shooting—the worst in US history—did however cost the premises owners dearly In 2017, 58 people attending a music festival in Las Vegas were shot dead and many more were injured by Stephen Paddock, a 64-year-old loner, who mounted his attack from the 32nd floor of the MGM-operated Mandalay Bay Hotel. Relatives of the victims brought a case against MGM claiming that it was negligent because it failed to monitor the hotel premises adequately. Paddock had stayed in the hotel for five days before the attack, stockpiling weapons and ammunition, using a "do not disturb sign" to prevent hotel staff from entering to clean his room.

MGM attempted to use the 2002 Safety Act as a defence. To give it its full name, it is the Support Anti-Terrorism by Fostering Effective Technologies Act, a piece of post-9/11 legislation. It was designed to protect companies that manufacture security products or deliver security services used to counter terrorism. It aimed to encourage innovation whilst capping liability should a product or service fail in the event of a terrorist attack.

The Las Vegas shooting was to be the first test of the Act, but eventually, MGM went for a mediated settlement. The 4,200 claimants, including families of those killed, injured, and traumatised, shared a $800 million pay-out, of which MGM's insurers covered $750 million. In the end, although the episode did not clarify the issue of who was liable in the event of a mass shooting, MGM's settlement suggests that they had little confidence in their defence.

Fulfilling a duty of care was getting more difficult, especially for low-probability, high-impact events. In essence, a property owner was likely to face massive costs should a guest be harmed whatever the circumstances.

To reduce the financial and legal exposure, owners therefore used three strategies. First, they increased security measures, including baggage screening, training staff to spot suspicious behaviour, and planning how to respond to active shooter events. Second, they transferred some of the risk of litigation by using external security providers, anticipating that they would be liable for security failures. And third, to hedge against financial exposure, they took out additional insurance. Following the MGM attack, the cost of liability insurance increased by 25%. And the "do not disturb" sign was no longer guarantee of a lie in.

The US is the world's most litigious country and lawyers became key players in the US security business. The lawsuit rate is four times higher than Canada, and almost twice as high as the UK. It also has a strong entrepreneurial culture which enables a *blame and claim* business to thrive in the form of ambulance chasing, *no-win no-fee* lawyers. This means that claimants, who might not be able to afford access to justice, can engage a lawyer at no cost, and that lawyers have a strong incentive to be aggressive. It also reflects the fact that money is the only means of compensating for harm suffered, and this is especially important in a society without state medical care. The consequence is a new breed of lawyers specialising in security cases. Their appearance on the scene also benefits security companies that can leverage examples of lost negligence cases to sell more services.

UK: Cricket and Canoes

The US and UK legal codes have common historical roots. Although most of the principles are the same, time, culture, and circumstances have led to different judgments on duty of care issues. The types of cases that define duty of care are somewhat different too. In the US, as we have seen, they feature mass shootings and terrorist attacks; in the UK, they feature ginger beer and snails, cricket and canoes. Wallace and Gromit to the US's Batman and Robin.

In the UK, the first successful negligence case was in Paisley, Scotland in 1928, where May Donoghue was bought a bottle of ginger beer by a friend. After drinking half its contents she discovered a decomposed snail in the bottle. She experienced stomach pains and visited a doctor, who diagnosed gastroenteritis and shock. She argued that the manufacturer, a Mr Stephenson, had negligently supplied a contaminated product. The Mr Stephenson argued that he had no contractual relationship with Mrs Donoghue as she had not bought the ginger beer herself.

The case went to the High Court, which awarded her £200 (worth about £13,000 in 2020). The judgment was based on the principle that, "You must take reasonable care to avoid acts or omissions which you can reasonably foresee, would be likely to injure your neighbour." It didn't matter that Donoghue had not bought the ginger beer herself; she should have reasonably expected it to be snail free.

The snail in the bottle case broadened the scope of liability, but the *reasonable test* remained nebulous, rather than mathematical. In 1947, at the Cheetham cricket ground near Manchester, a batsman hit a ball for six. It flew outside the ground hitting local resident Miss Stone. Cricketers dream of hitting a ball outside the ground, but it is a rare event. At Cheetham it had happened only six times in the previous 30 years.

The case went to the House of Lords who dismissed it ruling that the risk was *not reasonably foreseeable*. The Lords reasoned that if the law decided that every rare event was caused by negligence, nobody would do anything for fear of the remote possibility of an accident. It was a pragmatic judgment, but you can't help wonder if it was coloured by the Lords' fondness for cricket.

A step change in approaches to duty of care occurred in 2007 with the introduction of the Corporate Manslaughter and Corporate Homicide Act. This was designed to make it easier to convict the senior management of organisations that failed in their duty of care. It strengthened previous corporate manslaughter legislation, that had resulted in only a single conviction: the owner of an adventure centre who was jailed for negligence after serious safety breaches led to the deaths of four teenagers while on an organised canoeing trip in Lyme Bay in 1993.

Attempts to hold senior management to account in other cases had failed because corporate structures were so complex that pinning liability—and in the case of the criminal law, guilt—on individual managers proved impossible. The 2007 Act put senior managers firmly in the legal crosshairs with the threat of a large fine and a lengthy prison sentence. This was getting serious and it caught the attention of boardrooms everywhere. From being an obscure legal formulation, failing in a duty of care became a clear and present risk.

List the Risk

Security managers were used to making imprecise decisions based on their personal experience, sucking their teeth, and sticking wet fingers in the air. But taking *reasonable measures against reasonably foreseeable* harm demanded a much more methodical approach. This meant evaluating threats, listing ways to mitigate them, training staff, signposting hazards, monitoring the security environment, thickening up the relationship between those with safety and those with security responsibilities, and establishing formal chains of accountability.

A major step forward was risk assessments based on formal process rather than gut instinct. These had been in use in aviation and engineering since the 1960s where safety lapses in complex systems could lead to catastrophic events. The concept was embraced in health and safety from the 1970s, but it wasn't until the 2000s that it became common in the security world.

The key tool was a risk matrix, a grid showing the likelihood and impact of any given security event, from the loss of an ID card, to the detonation of a car bomb. The matrix normally had five levels of likelihood, from imminent to improbable, and five levels of impact, from catastrophic to minimal. Each level was represented by a colour, starting with innocent white, then green, orange, red, and menacing black for the highest risk. Suddenly security managers were aligned, not only with lawyers, but also with engineers. Their craft had a new-found precision with spreadsheets replacing wet fingers.

But risk assessment itself wasn't without peril, as a case in Italy highlighted. In 2009 the town of L'Aquila in Italy was struck by an earthquake that killed 309 people. In a hugely controversial decision, six Italian scientists and a government official were charged with manslaughter for providing citizens of the town with what the prosecution claimed was a falsely reassuring risk assessment. There was shock throughout the scientific community when they were found guilty and jailed for six years. The scientists were eventually acquitted, but the official, the deputy head of Italy's civil protection department, remained convicted with a reduced jail term of two years.

The case made everyone assessing risk more thorough, and more inclined to throw in handfuls of prudence. There were few rewards for calibrating risks, but avoiding jail was a big incentive for inflating them. A similar phenomenon occurred in health care, where increased litigation against doctors led to a practice known as defensive medicine. This is where extensive diagnostic tests and treatments are recommended, or doctors refuse to attempt risky treatments, primarily to protect themselves from potentially litigious patients. This results in extra process and increased professional liability insurance, without any improvement in the health of patients.

Roaming Risk

One of the biggest impacts of the widening interpretation of duty of care was on international work travel. Until the 2000s organisations would dispatch staff around the world relying on their good sense to keep out of trouble. That was the way it had always been. Columbus, Cook and Livingstone relied on their wits rather than their lawyers to come home safe. I hesitate to mention myself in the same paragraph, but I will anyway. In the early 1990s I cleared landmines for an NGO in Cambodia. The pre-departure briefing was a chat over a gin and tonic, we had no insurance, and there was little by way of medical support, communications, or security. We had staff held hostage, shot, robbed at gunpoint, killed and injured by landmines, and in road crashes. At the time it felt strangely normal, part of the adventure. We were all ex-soldiers and we dealt with situations as they occurred. No one complained, no one mentioned the phrase *duty of care*, and no one called a lawyer—even when staff lost a limb or their life. I'm not suggesting that this was good or bad, it was simply the norm for the times.

Fast forward 20 years and I was working for another NGO that delivered humanitarian aid in developing countries. Before any travel, staff were obliged to complete a 22-page risk assessment covering security, transport, safety, health, environmental, climatic, and safeguarding risks. It included detailed plans for security, medical support, communications, transport and accommodation and it had to be agreed and signed off by four managers at progressively more senior levels. It took a week to complete and was a stern test of administrative process and traveller commitment.

It was clear that over the course of two decades attitudes to security had profoundly shifted and that duty of care was a major factor. The changes were best summed up in a cartoon that circulated in the security community a few years ago showing a father reading a bedtime story to his sleepy child with the caption, "I can't believe that Goldilocks went into the woods without a risk assessment!"

Yet the debate about what is *reasonable* continues. When I was posted to the British Embassy in Myanmar in 2015/16, diplomats used official vehicles for work travel. These were normally Landrover Discoveries that had been imported from the UK complete with seat belts and airbags. Seat belts were not mandatory in Myanmar and the official vehicles were only available for work. Some staff expected UK standards of road safety arguing that, to fulfil its duty of care, the embassy should run a service with cars fitted with seatbelts available for their personal travel in the evenings and at weekends.

The embassy held firm pointing out that staff had volunteered for a posting to Myanmar having been informed that embassy vehicles were not available for their private travel. If staff wanted a car for personal use, it was their responsibility to identify one with seat belts.

Whilst I was in Myanmar, no embassy staff were injured for want of a seat belt. But had they been, the outcome of any resulting litigation would have been hard to predict. If the Foreign Office had lost the case, it may have been obliged to provide vehicles with UK safety standards for off-duty staff at many of its 165 embassies.

Even trees have been drawn into the duty of care debate. Bristol Council was ridiculed in a 2006 Daily Mail article under the title, "Yew Couldn't Make It Up! Nanny Council Chops Down 100 Yew Trees Next to a Playground in Case the Children Poison Themselves by Eating the Leaves." A Council risk assessment had concluded that if children in the playground ate sufficient quantities of leaves, they might vomit. The case was easy to mock, but it demonstrated that fear of litigation and desire to avoid harm, had become a major preoccupation.

The following year, trees again featured in a duty of care case, this time in tragic circumstances. Eleven-year-old Daniel Mullinger was killed by a falling branch on National Trust land in Norfolk. His parents claimed that the National Trust was negligent because a routine inspection did not identify the weakness in the tree. The judge recognised that it was "the cruellest coincidence" that the branch fell as Daniel passed beneath it. He took into consideration the remote chance of a branch falling and the low number of people using the path beneath the tree. His conclusion was that to avoid the tragedy the National Trust would be required, "to do more than was reasonable to see that the children enjoying the use of this wood were reasonably safe." In other words, it was disproportionately expensive to maintain trees to reduce the already remote risk of a tragedy.

Whilst there was no compensation in this case, it influenced the behaviour of all parties. Landowners became more alert to their possible liability for falling branches and paid closer attention to risk assessments. Insurance companies increased liability premiums. Tree maintenance companies became more conservative in their judgements and harsher in their pruning. Parents became less willing to allow their children to play in nature. Everyone became more risk-averse which reduced the likelihood of harm. So even unsuccessful cases shape attitudes to safety and security.

Duty of care also provided a platform for private security companies to provide additional services. For example, International SOS, a major travel health and security company, produced a white paper in 2009, on how to deliver robust duty of care measures listing dozens of areas of best practice from crisis management plans to traveller tracking systems. The days of staff just buying a plane ticket, booking a hotel, and muddling through were over. You had to be assessed, briefed, trained, tracked, submit proof of life questions in case you were kidnapped, have a medical examination, stay in security-approved hotels, only use company-authorised transport, be provided with alerts if the security situation changed, and sign to acknowledge that all points have been understood. It was exhausting before you even got on the plane.

In 2016, a sign that duty of care has come of age was the inaugural Duty of Care Summit and Awards. This brought together safety and security providers and practitioners to share best practice and to recognise innovation and leadership in duty of care, which they defined as "the moral and legal obligations that employers have to their workforce to maintain their wellbeing, security and safety at home and abroad." It was no coincidence that the event was organised by International SOS and that the major sponsors of the event were the insurance companies Marsh and Chubb. Naturally, there was a commercial aspect to it, but it demonstrated that duty of care was no longer an onerous obligation, it was a virtue that made a workforce happier, healthier, and more productive.

Artificial Lawyers

Law is an ancient profession, but it is feeling the impact of one of the newest of technologies—artificial intelligence—or AI as it is widely known. Business information firm Thompson Reuters estimates that in the US alone, expenditure on AI legal solutions will increase from $12 billion in 2017 to $85 billion in 2027. AI is brilliant at searching for information, answering legal queries, document management, and generating and analysing contracts.

AI now powers online claims portals which reduce processing times from months to days, saving all parties a lot of money along the way. An in-house program was developed by financial services firm JP Morgan. Called COIN, short for Contract Intelligence, it can do the dreary job of reviewing commercial contracts in double quick time, and saves the company 360,000 hours of lawyers' time each year.

Lawyers, like retail checkout staff, will see their professional opportunities decline as the AI revolution picks up pace. A lawyer's judgment will remain important for complex situations, but online systems will become the norm for straightforward cases. As awareness of duty of care grows, costs of making a case will fall, and the speed of resolution will increase. As a result, people will be more inclined to pursue a claim. So, we can expect to see an acceleration in the number of duty of care cases, and a further disinclination to take risks.

Duty of care has come a long way since Mr Winterbottom and his broken coach wheel, continually pushing up standards, redefining policy and generating new expectations. The journey reflects the development of Anglo-Saxon law, history, economics, societal norms, safety, and compensation culture. It's a collective endeavour, the result of thousands of lawsuits. There is no central plan that identifies what security policies need to be challenged. No individual or organisation is in control. It is an organic process driven by new situations being tested, and assumptions being made about how to define it.

There are many positives to duty of care. It holds people to account and compensates for harm. It provides a comprehensive means of considering risk. It's also a two-way street; employers have a duty of care, but employees have a duty to comply with safety and security regimes, so it is a shared responsibility. And it signals virtue, demonstrating that an organisation really does care for its staff, which helps to attract talent.

But it is not without downsides. It turns security management from an instinctive process, into an endless compliance process designed to avoid both injury and litigation. It promotes an abundance of caution and blame-seeking, whilst inhibiting risk taking. It escalates the cost of everything from delivering aid, to car parking charges, to staying in a hotel, to insurance cover. It creates what Professor of Risk, John Adams, calls "Compulsive Risk Assessment Psychosis," (sometimes known by its acronym) where common sense is replaced by bureaucratic process. And it continuously forces higher standards with no obvious end point.

Duty of care will be an increasingly influential factor in shaping security and the debate about positives and negatives will continue; as US civil rights lawyer Charles Hamilton Houston said, "A lawyer's either a social engineer or … a parasite on society."

13

HEAVEN OR HAL

The richest companies in the world are the US tech giants Apple, Amazon, Microsoft, Google and Facebook. Their influence and dominance have a similar level of impact on the 21st century that Genghis Khan had on the 12th century. Genghis grew his empire through military conquest, while tech giants grow theirs through digital conquest. Like all Emperors, tech bosses—the likes of Steve Jobs, Bill Gates, Elon Musk, Jeff Bezos and Mark Zuckerberg—have an unswerving belief in the goodness of their own cause. Like Genghis, they also believe that if everyone submitted to them, everyone's lives would be richer and more fulfilling, and the world would be a better place. They have no shortage of ambition, righteousness, or resources as they advance with missionary zeal.

Tech giants have created new paradigms for how we access information, interact with each other, communicate, and buy things. They also set new standards for how they manage their staff. To attract talent, they aim to offer the best salaries, and to retain them, they provide the best perks, the best working environment, and the best security.

The Valley of the Tech-Kings

All organisations assimilate and project the culture of their headquarters' location. This includes the tech giants which are mostly based in California's Silicon Valley. California, "the Golden State", represents freedom, liberty, diversity, optimism, confidence, health, modernity, and cool. And it's rich. If California was a country, it would have the 5th largest economy in the world.

Once famous for grapes and oranges, agriculture in Silicon Valley was gradually displaced, first by Stanford University in the 1880s, then by defence contractors in the 1930s and 40s, then by electronics companies in the 60s, personal computer companies in the 80s, and by tech companies in the 90s. By the early 2000s labs and geeks had entirely replaced fields and farmers.

Stanford grew in status to become one of the world's major academic institutions, setting out the West Coast's challenge to the East Coast's Ivy League and developing a reputation for entrepreneurship and nurturing start-ups. During World War II ships, fighting vehicles and aircraft were manufactured in the Valley. Defence contractors remained there to supply the US military with increasingly sophisticated equipment during the arms race of the Cold War. The combination of entrepreneurial academia and large-scale manufacturing, drew in capital to fund technology projects, nurturing innovators such as Hewlett Packard, Lockheed, IBM, and Xerox. Firms started making transistors, integrated circuits, and semiconductors for use in the new electronics industries. From there it was a short step to computers, software, and the internet. It soon had the highest concentration of tech jobs in the world.

But financial success comes at a price. The phrase "Silicon Valley" may conjure images of bucolic landscapes, fruit trees and tasteful modern buildings nesting among gently rolling hills. The reality is a vast urban dystopia divided by mega highways. The tech giants' explosive expansion places extraordinary stress on the surrounding infrastructure. Between 2015 and 2018, traffic speeds halved, while real estate prices doubled. Meanwhile, the sight of destitute people sleeping on the streets tarnishes California's golden reputation.

Tech Life

Life for the tech worker means living in overpriced, cramped accommodation, commuting in dense, slow-flowing traffic, while sucking flavoured coffee from an insulated mug and making hands-free phone calls, working in massive open-plan offices, shopping online with Alexa, and having Mexican food delivered to their door each night. As most tech workers move in from out of state, they don't know anyone in the area, so many turn to dating apps for human interaction outside work. It's a curiously dysfunctional impersonal environment from which to incubate the world's largest social networks.

I worked as a Security Manager in Facebook's physical security team, known as Global Security, with a mission to, "protect people, assets and reputation".

Facebook, in the words of its executives, "is one of the most consequential companies ever". They have a point: almost half of all people on earth use its services (Facebook, WhatsApp, Instagram, Messenger). Mark Zuckerberg founded the company in 2004 and by 2021 it had revenues of $120 billion. Not content with creating the world's most successful social networks, he was on the cusp of launching a crypto currency with the modest ambition of transforming the world's banking. The company also developed a dating service that, presumably, will transform the way we reproduce. And, not content with conquering the real world, Mark (as he likes to be known inside the company) has now created a virtual world known as the Metaverse. Its influence, speed of growth, and ambition is breathtaking.

Around 90,000 plump (I'll explain in a moment!) millennials worked in Facebook's Silicon Valley headquarters before, in Mark's words, he had to let go of 11,000 "talented employees" in November 2022. The headquarters is an immense series of adjoining university-style campuses, each so big that you could park an airship inside. These are the crucibles of the 4th Industrial Revolution. But instead of the howl of the steam engine and the clank of hammers, there is the woosh of the espresso machine and the hum of the hard drive. Like sweatshops of old, everyone works in tightly packed rows 120cm apart, but the looms and presses have been swapped for elevating desks and ergonomic keyboards. Instead of grimy overalls and weary frowns, everyone wears hoodies and perma-smiles.

The walls are brightly coloured and pasted, not with thoughts from Chairman Mao's *Little Red Book*, but with the words of CEO Mark's big red company values: build social value, be open, focus on impact, be bold, and move fast. The ceilings are exposed, revealing pipes, beams, cabling, smoke detectors, sprinklers, and bare concrete. This deliberately raw feel is consistent with another of Mark's slogans: "This journey is 1% finished". By my reckoning, if $120 billion is 1%, he'll be reaching his peak at around $1,200 billion. And I don't think he is joking.

The thoughts of Chairman Mark.

The offices combine fun with efficiency. There are beanbags, massage chairs, board games, music rooms, pool tables, play stations, pinball machines, sleep pods, and rooftop gardens. The offices are designed so that the thousands of staff can reach any room from anywhere within 3 minutes. Facebook meetings are organised in 30-minute chunks, starting exactly on or half past the hour. So, every 30 minutes the lifts and corridors are filled with millennials, headphones on, Mac open in one hand, and a cappuccino in the other, dashing between rooms. Punctuality in Facebook rivals that of Swiss railways.

Fostering Contentment: The Facebook 15

All this frenetic activity needs fuel and Facebook supplies that in gargantuan quantities. All offices have restaurants offering breakfast, lunch, and dinner, absolutely gratis. Organic scrambled eggs, grilled salmon, steamed asparagus, gluten-free pizza, ice cream, fresh juice, pancakes, lark's tongue in aspic…. There is no hunting around the neighbourhood in the rain for an egg sandwich. Teams eat together and "build social value". And everyone is nourished and back at their desks within 30 minutes. It's brilliant.

But there's more! At every turn there are groaning racks of biscuits, chocolate, crisps, granola bars, vitamin drinks, soda, yoghurt, gum, nuts, liquorice, and even the odd bit of fruit. There is also beer, tequila, and Prosecco on tap—all free! Millennials are more inclined to take Adderall than alcohol, but they are amused to see boomer booze on display.

Facebook keeps you fat, happy, and focused, nourished in its ample bosom. There is a phenomenon known as the "Facebook 15" because, it is said, everyone puts on 15 pounds within 3 months of starting at the company. But there is serious intent behind the food and the fun: Facebook wants maximum productivity from its staff.

For instance, as well as catering for all your nutritional needs, there is valet parking. If you haven't yet gone electric, there's a fuel truck that will top up your car while you are at work. Everyone is given their own smartphone and computer which are upgraded every year. On campus you'll also find a dry-cleaning service, a gym, a hairdresser, and a masseuse. You can get your Amazon package delivered to work, invite your friends to lunch, use vending machines to get anything from flash drives to batteries, Bluetooth headphones to thank-you notes—all free. It's a utopian vision of benevolent capitalism. The company takes care of your time-consuming personal administration, freeing you to focus on work.

Chapter 13: Heaven or HAL **193**

Free food! Facebook keeps everyone fat, happy and focussed.

As well as feeding, laundry, refuelling, and hairstyling, Facebook also takes care of staff security, on campus, at home and while travelling. Global Security treats staff as valued customers. Once part of the Facebook family, in the same way that you don't have to worry about feeding yourself, you don't have to worry about security.

Staff can seek advice, information, and receive security assessments 24/7. When flying outside the USA or Europe, employees are met airside, chauffeured in executive vehicles, accompanied by a bodyguard and accommodated in the finest hotels. In an attempt to recreate Silicon Valley levels of security, a travelling Facebook software engineer is provided with more protection than most US diplomats. Facebook staff are expected to negotiate the cyberworld independently, but in the real world they need specialist help.

Facebook uses its own products for internal communications, so everyone's attention span is shredded. In meetings, people occasionally glance up at whoever is speaking, but most of their time is spent flipping constantly between WhatsApp, Workplace and Workchat (the commercial versions of Facebook), Messenger, Instagram, SMS, and email. Just like teenagers looking for a hook-up on a Saturday night.

This devotion to screen time may be second nature for a millennial, but I punched out my undergraduate thesis on a ribbon typewriter, got my first mobile phone in my 30s, and I still use two fingers to type. Despite wearing jeans and a hipster beard, I sometimes felt like an analogue man in a digital world.

I soon discovered what happened when Facebook's all-embracing security service combined was with a millennial's sense of entitlement. I received a call early one morning from a junior marketing assistant who was put up in a fashionable Soho hotel during her work trip to London. "I need your help," she demanded. "When I woke up, I took a drink from a bottle of water on my bedside table. It was contaminated! I could be seriously ill." I was immediately alert: rather than call the operations centre, she had used the internal directory to identify me as the most senior security person in the region. A millennial full of indignation and dodgy water—this was indeed a crisis!

I held back my analogue instincts, put on my concerned voice, and asked how she was feeling. "I might be ok, but here's what I need," she rattled off a list, "a medical examination, a laboratory test on the water, somewhere else to stay, a lawyer to put in a civil claim, and you need to blacklist this hotel." I knew that this was a career-defining moment. As a Boomer in Facebook, I felt like I was in Pol Pot's Cambodia, as depicted in the film *The Killing Fields*, where being denounced by a child could lead to instant death.

I put on my *oh no, how terrible voice* and offered my sympathy. I then asked if the bottle of water had been sealed. "I'm pretty sure it was," she said. Her reply set off alarm bells but fearing that further questioning might lead to a shallow grave, I initiated all her requests. An hour later, she called, and said with a remarkable absence of sheepishness, "I just remembered, I filled the bottle in the hotel's gym and the dispenser had lemons in it…." I managed to take a deep breath and say, "as long as you are ok, is there anything else we can help with?" hoping she couldn't hear my eyes rolling to the back of my skull.

Silicon Pally

Facebookers greet you warmly: they ask about your family, are considerate and polite, look you in the eye when they talk to you, have good teeth, smile a lot—and they smell nice. Apart from all that, I have nothing against them.

They also have a missionary earnestness about them. You sense Thought Police gauging if you are a true believer and fully embracing the programme, as they make eye contact that little bit too long.

Facebook is a tech company so everything happens fast. See what happened to Blackberry when it didn't move fast enough? And whilst the mission is to "give people the power to build community and bring the world closer together", it has a relentless drive to become a gazillion-dollar company.

At an annual sales meeting in a cavernous venue outside Dublin, I saw the European boss, Nicola Mendelsohn, deliver an inspiring speech to 3,000 staff ending with the words: "Our aim is, growth, Growth, GROWTH!" Her outstretched arms lifting to the heavens as the hall erupted with claps, cheers, and whistles, increasingly louder with each successive "growth". It was a Hallelujah moment for secular commercialism.

Facebook is informal but up-tight, fun but serious, open but guarded. And it has a level of internal compliance that many armies would envy. The five core values shape the culture, but it doesn't stop it feeling like *Game of Thrones*, as everyone jockeys for position as the company scales and multiplies. The "focus on impact" value, combined with a six-monthly individual performance cycle, keeps everyone scratching for "wins".

The aim of security in any organisation is to ensure that no incidents occur, but you can't score a win by nothing happening. In Global Security there is a saying, "one 'oh shit', wipes out 100 'ataboys' ". So you are in a constant state of anxiety whilst checking that nothing happens. A win is therefore a contribution to policy, the successful opening of a new office and especially, positive feedback from a senior member of staff.

In my first month, a senior executive in the London office asked me for some security advice. I responded quickly, had what I thought was a productive meeting and was thanked for my service. I thought I had achieved a win, but the following day I received a request for an online meeting with a senior member of the Global Security intelligence team based in Silicon Valley (who until recently had been a junior police officer). I was expecting to be applauded for my sound advice. Wrong! I was subjected to a 40 minute head scrubbing for speaking to an executive when I didn't "own the relationship." The only saving grace was that it was on-line so his spittle dribbled down the screen rather than my face. How foolish of me not to know that that particular win should have gone to him rather than me.

Global Security

The gods of the company are the software engineers (the Geeks are inheriting the Earth!), who make the on-screen magic happen. Then there are the sales and marketing staff, who sell the $120 billion worth of adverts that pay for all those candy bars, and then the Community Operations teams, the folks that take down endless amounts of inappropriate content. Then there are the support services like HR, Finance, Facilities, Culinary, and Security.

In 2015 Global Security was a gang of about 10, unfettered by substantial experience or self-doubt. Its structure now resembles that of an army, with more than a dozen grades from Door Security Officer to Chief Security Officer. This is more than twice as many grades on the Catholic Church's career ladder from priest to Pope, and pretty impressive for an organisation that prides itself on its flat structure.

One of the key tasks is protecting Mark. Until 2012 he used to drive himself to work in a modest Honda. But this was unsustainable as the company's profile and fortunes multiplied. Being the CEO, founder, and the very public face of Facebook, he made many friends. But he, and his company, were not universally popular. Controversies included defamatory content, fake news, suspended and closed accounts, data protection, and influencing the outcome of elections. All this led to hate mail and death threats.

Normally hate mail is an unpleasant nuisance and threats remain just that. If you really intend to harm someone, you would not advertise it in advance. These *fixated persons* are, like wolves, divided into *howlers* and *hunters*. The ones who make threats are known as howlers. They are noisy, but seldom dangerous. The dangerous ones are the *hunters*, who don't draw attention to themselves before striking.

Mark would receive hundreds of thousands of threatening messages every year. If there was just one hunter for every thousand howlers, there was a vulnerability that needed to be reduced. If Mark were killed, billions would be wiped off the share price of the company.

The price of his fame and wealth was submitting to 24/7 security. It sounds glamorous, but having body guards is highly intrusive. In the world of diplomatic security, ambassadors with dedicated close protection are given an additional two weeks a year of home leave to compensate for having a bunch of muscly men and women share their lives. Tech billionaires have many privileges, but privacy is not one of them.

Mark has more bodyguards than Madonna, and only slightly fewer than the US President. Facebook's 2019 financial filings revealed that his security, which is headed by a former Secret Service agent, cost $24 million a year (which is about the same as the entire defence budget for Moldova). On top of that, his office has levels of bomb protection not seen outside Baghdad's Green Zone,

Chapter 13: Heaven or HAL **197**

and is rumoured to have a secret passage to spirit him away him should there be an incident.

Facebook's European Headquarters is the low tax haven of Dublin where they took over the site of the Allied Irish Bank in 2018 and transformed it into offices. The bank had, for many years, stored massive cash reserves, yet the security measures were considered inadequate for Facebook and its rows of computers and culinary treats. Small wonder that in 2019, Facebook's physical security budget for offices in 30 high-income countries rivalled that of the United Nations' operations in 150 countries, including 20 war zones.

Facebook's hard security measures have a soft exterior. The aim is to give staff and visitors a "frictionless, white-glove service" and security is blended with customer service to provide a Disney-style *visitor experience*. Around Facebook offices you'll often see cute black Labradors, known as "Paws on Patrol." They are not trained to attack or detect anything; they are comfort dogs, who, like all warm-blooded creatures, enjoy being stroked and their main purpose is to project a cuddly image of security.

As with many companies, the visible security staff, who guard offices and manage access control, are contractors. With a lot of security, the emphasis is on big muscles. At Facebook, it is on big smiles. Security officers must provide cheery greetings at a moment's notice: "Happy Monday and welcome to Facebook!" You do need to be unrelentingly cheerful to work in these roles.

Welcome to Facebook: Security blended with customer service.

Retention is a major issue in the security sector. Most companies would be delighted to retain 50% of their staff annually, but Facebook regularly retains more than 80%. Security contractors value the buzzy environment and the lavish operation, and they nurture the hope of one day becoming full-time employees. And of course, the free food is a major perk for these lower-paid workers.

Central Control

Most billion-dollar companies have headquarters corporate security teams of perhaps 20. The UN has a team of about 100 in its New York HQ, managing the security of 400,000 staff worldwide in the toughest of environments. Facebook has a HQ Global Security team of more than 350 with a further 7,000 contract staff at offices in the US and worldwide.

HQ Global Security has five directorates (Operations, Intelligence, Strategic Initiatives, Security Services and Executive Protection) which are themselves split into more than 30 separate teams with names like Operational Excellence, Strategic Business Partnerships, and Protective Intelligence. Then there are uniformed staff working in darkened control centres that look as if they could coordinate a moon landing. Screens glow with CCTV images, news feeds, performance data, maps, and travel information. Operators with headphones chatter into microphones as their eyes flicker across their multiscreen workstations.

The HQ structure is replicated on a smaller scale in regional centres around the world, with all the various teams reporting to HQ. New teams are spawned monthly making internal alignment ever more complex. It's like being in a crowded bed and having to move over every few minutes to make room for new people. You don't stay comfortable for long, and, like readers of the *Karma Sutra*, you need a big appetite for new positions.

The emphasis is on collaboration rather than coordination in a highly centralised system that is effective only because of the sheer weight of resources. Counterintuitively, as the company expands, individuals' responsibilities contract. Everyone becomes a hyper-specialist like a worker on a production line. The mantra is "stay within your swim lane" even as they become ever narrower. The vast numbers of staff, the lack of coordination and obsession with "wins" makes it feel like a game of school yard football with everyone swarming around trying to get the ball rather than passing it to each other.

But behind the smiles there is the system, ready to respond to a range of circumstances at a moment's notice including demonstrations, unattended bags, power cuts, bad weather, lost ID cards and suspicious persons. Everything is reported to a SOC, from where guards are talked through the standard operating procedure playbook via an ear phone connected to a radio feed.

Chapter 13: Heaven or HAL **199**

Everything under control at a Facebook SOC.

The system can be very effective, but procedures don't always allow for common sense or human empathy. At the main London office, a woman and her disabled daughter arrived at reception to complain that her Facebook account had been hacked. She was in some distress and in an exasperated voice said, "if I can't get this fixed, I'm going to kill someone." The security officer was obliged to report the incident to the control room, saying it was just a figure of speech and he had the situation under control. The playbook determined that as it was categorised as a "violent threat" it had to be escalated to the security control room in the US which insisted that the police must be called.

Meanwhile, her daughter had soiled herself, adding to the woman's distress. The security officer showed them the restroom, and had a colleague go out to buy diapers. Once they were cleaned up, he gave them a seat, made them a cup of tea, and had another colleague talk them through the online help procedures. They eventually left in a much better state of mind.

The internal recriminations started immediately. "Lives had been put at risk… trust had been eroded… it was irresponsible… why had procedures not been followed?" We explained that, had the police been called to remove a distressed woman and her disabled daughter, it could have escalated into a reputational disaster. Tea rather than truncheons was a better way of resolving the situation.

Global Security was then headed by Nick Lovrien, the Chief Security Officer, and a man who knows the difference between a clarinet and an oboe. Having started his professional career with the electronics retailer Target, he spent five years with the CIA before returning to Target and then moving to Facebook in 2013, when there were only a handful of people in the security team.

On the Facebook app, if you don't post a photograph of yourself, a silhouette of a person's head appears. It is of indeterminate gender and the hair forms a pronounced curled spike above the head. I was never sure if Nick modelled his hair on the silhouette, or if the silhouette was modelled on him.

At the regular Global Security meetings the key thing was to demonstrate that you had drunk the Kool-Aid. The speakers often displayed more enigma than charisma, as they recited lists of values, priorities, objectives, and goals, while everyone listened with keen eyed attentiveness, occasionally bursting into applause or letting out whoops of "awesome". It was like Alexa reading a shopping list in an evangelist church.

Senior managers from Silicon Valley had an unnerving habit of appearing unannounced at their subordinates' desks in offices in Europe or Asia, smiling, cup of coffee in hand, as if they were just passing by. "Oh hi, wassup? great to see you!" they would say as you'd try to replace the surprise on your face with something more welcoming. It felt more like a snap room inspection by prison guards than the visit of an executive promoting openness and social value.

BOLO

Between 2015 and 2018, Facebook transmogrified from David to Goliath. And the trust that arrived on foot departed on horseback, with user data in its saddlebags. But that doesn't necessarily translate into a loss in revenue, or real-world security threats. The main issues are users with account problems, haters that have their content removed, protestors concerned about digital colonisation, or the minuscule taxes paid on the massive income. All of this is modest compared with the daily threats facing transport networks, media organisations, retail empires, or hotel chains. Let alone the UN or governments with embassies in conflict zones.

Global Security keeps data on suspicious people: those who have posted inappropriate content on the Facebook app, or those who are hostile to the company. Classified as "BOLOs"—BOLO stands for "be on the look-out"—they are forbidden to enter Facebook premises. The concerning aspect of this is that Facebook has access to BOLO data: where they live, who their friends are, how they communicate, what their preferences are, their holiday locations, their politics. It must be very tempting for Facebook to use that information to monitor BOLOs, especially if they are known to have issued a threat to the CEO.

Threats to tech companies are not entirely overblown. In April 2018, for example, a disgruntled user shot dead three people at Google-owned YouTube's Silicon Valley office. There was immediate concern that one of Facebook's 2 billion users might take a similar route. So, dozens of off-duty police officers were enlisted to discretely patrol Facebook's offices. Inevitably a game of "spot the cop" commenced. A straight-backed middle-aged man with a shaved head filling his pockets at the snack stand was normally quite conspicuous amongst the slouching goatee-wearing millennials.

If you join Global Security from the police or military, it can feel like a parallel universe as you recalibrate from handling life-threatening situations to dealing with elevator entrapments and queues of excited visitors. It should be a joyous transition, but it feels highly strung and humourless. Information is held so tightly within each narrow speciality that no one has full visibility of what is going on. The Mafia would be impressed by the adherence to the "need to know" principle. However, even Facebook's junior security managers are paid more than senior doctors, so they are content to be clasped in golden handcuffs, get with the programme, and let out their belt buckles regularly.

HAL

Facebook manages modest threats with big cash and high anxiety. This is coupled with an ambition that would make Napoleon look timid and an "abundance of caution" approach that makes nuns seem reckless. Most organisations manage risk, carefully evaluating impact and likelihood, and balancing resources against possible losses. At Facebook the emphasis is on controlling everything all the time, whilst smiling broadly.

Its security is intelligence-led, process-orientated, and supported by automation that aims to be consistent, replicable, and scalable. And despite the legions of staff, it feels like Global Security is advancing towards an artificial-intelligence, algorithm-based system that one day, like driverless cars, will dispense with humans. Algorithms are of course just opinions embedded in code. Those opinions are based on the judgement of millennials whose greatest challenge is buying the latest lycra gym wear from Lululemon's in Stanford Mall. It's a narrow experience to act as the reference point for international security.

Much of the technology and technique underpinning all this is impressive, but it's also concerning. Accountability is unclear, judgement is subordinate to procedures, and I wonder if anyone understands the whole process. It's as if HAL 9000, Arthur C. Clark's sentient computer in *2001: A Space Odyssey*, is on the cusp of taking over Global Security.

The procedural approach, combined with worst-case-scenario thinking, can breed fear and irrationality. Sarin gas is a military-grade weapon of mass destruction, used on very rare occasions by Middle Eastern despots, and once, in 1995, by a Japanese cult on an underground train in Tokyo. But mail arriving at Facebook is scanned for all conceivable menace and in July 2019, the media reported that a scanner alerted for Sarin gas and thousands of staff were evacuated. It turned out to be a false alarm. The irony was that one of the company's largest-ever operational disruption events was caused by a security system that was designed to protect it.

Global Security sits at a unique confluence of an abundance of wealth, tech, ambition, and caution as it protects Facebook on its march towards global digital hegemony. While it may aspire to set the gold standard for corporate security, its approach doesn't project neatly onto the complex, pragmatic, and parsimonious real world. But there is no doubting that it represents something of a paradigm shift, and I was left wondering if it was heaven or HAL.

14

PHREAKS, GEEKS, AND HACKERS

The first time cybercriminals in the US netted more money than traditional bank robbers was in 2010. Bank security was getting tighter, the use of cash was in decline, the average robber got away with only $4,000, and there was a significant chance of getting shot in the process. By 2020 bank robbery was a fast-disappearing trade: FBI figures show there were only 1,788 hold-ups, down from 5,546 a decade earlier.

Robbing a bank is an adrenaline-fuelled, high-risk activity, that demands reckless courage. A cyber robbery, by contrast, is easier, unlikely to end in a prison or a mortuary, and can be pulled off by geeks, wearing pyjamas, in between rounds of *Call of Duty*. By 2018 it wasn't just banks, almost half of all Americans reported being victim of cybercrime.

This chapter looks at the rapidly growing cyber security industry and the protection of devices, networks, and data, from online attacks. It is the fastest growing part of the modern security business. Worldwide in 2004 the cyber security business was worth around $3.5 billion. By 2020 it was worth $184 billion. Growing at a rate of 10% a year, it will overtake the long-established, $250-billion physical security market by 2026.

"Cyber" refers to the electronic world of computers and interconnected digital technology. The term has curious origins. It has its roots in the Greek word *kubernētēs*, which means helmsman. It next appeared in 1948 in a book by US mathematician, Norbert Weiner. Titled *Cybernetics*, it discussed the concept of control and communication in the animal and the machine worlds. The term might have remained obscure had silver humanoids known as *Cybermen* not appeared in the British TV sci-fi series Dr Who in the 1960s and 70s, sending a generation of terrified youngsters scurrying behind the sofa.

In 1982 American writer William Gibson wrote a sci-fi novel called *Neuromancer* which described a data thief in a digital future and used the term *cyberspace*. Many users of early computers also had a fascination with sci-fi and it was this fertile combination that incubated the spread of the term *cyber*.

There are many reasons for the rapid growth in cyber security. Foremost is the expansion of the cyberspace. It started with a 1960s US Government project called ARPANET which networked computers so that they could talk to each other. From there it spread to universities where it was used by academics to share information. In 1990, Tim Berners-Lee, a British scientist working at the CERN (the European Nuclear Research Facility in Switzerland) expanded the network, enabling anyone to connect to anyone, anywhere. He called it the world wide web.

The world wide web was the most profound invention since the wheel. It was also the most prolific. It took 60 years for 50 million people to buy a car, 22 years for 50 million people to get a TV, but only seven years for 50 million people to get online. By 2005 there were a billion users. By 2020 close to 5 billion people, more than 60% of the world's population, were connected and accessing the sum of all human knowledge (and watching an awful lot of cat videos). Every aspect of life from banking, to booking a table at a restaurant, to checking the football results, was online. People in western countries now spend 7 hours a day, almost half their waking lives, staring at screens connected to the web, and the rest of the world is catching up fast.

The creation of this cyberspace inevitably led to new opportunities for crime. The British government estimated that in 2020 cybercrime was costing the UK economy £27 billion a year. The US Center for Strategic and International Studies claimed that it was costing the global economy close to $1 trillion a year, around 1% of global GDP. And it was rapidly accelerating, up 50% from 2018 to 2020. People may not have been murdered or physically assaulted in cyberspace, but they were being defrauded, their personal data and intellectual property was stolen, access to critical networks was being denied, and ransoms were being demanded. Cyber security had exploded into a major issue.

Cybercrime

There are four broad categories of people involved in cybercrime. Hackers, who infect systems for malicious pleasure. Activists, pushing a political or social agenda. Criminals, seeking financial gain either by stealing money, data, or intellectual property. And state-sponsored groups, who aim to steal state secrets, manipulate public opinion, or test military responses.

The story of cybercrime predates the creation of cyberspace. Telephones were the earliest electronic networked systems. In 1970 a call from London to New York cost 50 pence a minute. For 50 pence in 1970 you could buy two pints of beer and a packet of crisps. So, people looked for ways of getting cheaper calls. A loose community known as *phreaks* (phone—freaks) developed. They had an interest in electronics, and by subverting the telephone network, they managed to talk to distant friends without digging into their beer fund.

Some telephone companies used tones to signal when calls had ended. Phreaks discovered that a commonly used tone, 2600Hz, was produced by a whistle given away in boxes of *Capn' Crunch* breakfast cereal. From there, phreaks developed a device known as a Blue Box that enabled telephone fraud by manipulating billing.

The first digital version of the Blue Box was built by a couple of young, long-haired electronics enthusiasts called Steve Jobs and Steve Wozniek (who would later go on to found Apple computers). "Experiences like that taught us the power of ideas" said Steve Jobs in 1995, "If you could build this box, you could control hundreds of billions of dollars around the world, that's a powerful thing. If we wouldn't have made blue boxes, there would have been no Apple."

As computers became more popular in the 1980s, the phreaks with their interest in by-passing electronics systems, naturally progressed into hacking. They remained a fringe sub-culture until the 1983 release of the film *War Games* starring Matthew Broderick as David Lightfoot, a high school hacker who accessed a US military computer. What started as a mischievous prank ended up with young David preventing a nuclear war and saving the world. It turned hackers from zeros to heroes and motivated a new generation to join them.

War Games may have only been a movie with a far-fetched plot, and it came out years before most people had even seen a real computer. But the US government was sufficiently unnerved to pass the 1986 Computer Fraud and Abuse Act which outlawed unauthorised access to computer networks. The first person to be convicted under the Act was Robert Morris, a Cornel

University undergraduate who in 1988 designed a worm—a type of virus that replicates itself and spreads amongst computer networks. Worms and viruses are part of an unhappy family known as malware, as in malicious software. Other members of the malware family include Trojan Horses (that look legitimate but conceal a virus), spyware (that collects information), and ransomware (that denies access to a system until a ransom is paid).

Morris claimed that he was motivated "to demonstrate the inadequacies of current security measures on computer systems." He was more cyber-explorer than cybercriminal. But his worm had a bug of its own that sent it out of control and it infected thousands of computers, slowing down networks and causing millions of dollars' worth of damage. Morris was fined and sentenced to 400 hours of community service. He recovered from the episode and eventually become a professor of computer science at Harvard.

News of Morris's accidental malware spread, and before long people started to experiment and launch versions of their own. The history of cyber security is full of young men, self-taught and acting alone, who deliberately spread malware. One of the most notorious was David Lee Smith, an American computer programmer, who in 1999 spread a malicious virus known as *Melissa*. He posted a file on an adult internet site promising free passwords to feepaying websites. When users took the bait, a virus was unleashed which soon hit millions of computers. The FBI estimated that the disruption cost $80 million. Others include Filipino college student Onel de Guzman, whose *I Love You* virus infected more than 10 million computers in 2000, and German teenager, Sven Jashan, whose *Sasser* worm caused damage costing $500 million.

As malware attacks increased, two forms of cyber security measures became commercially common: firewalls, which prevented unauthorised access, and anti-viral software which identified and removed worms and viruses. A firewall is much like the physical structure that prevents fire from spreading in buildings. In some ways it is also like an army that takes care of external threats. Anti-virus software scans, detects and deletes viruses. Rather like the police, taking care of internal threats. But like all security measures, not only do they have to be installed in the first place, but they must also be regularly updated to be effective against constantly evolving threats.

In 1995 the internet's floodgates opened for commercial traffic after other key security developments made it safer to conduct financial transactions online. The first was the Secure Sockets Layer, or SSL, which encrypts data. The second was the Hypertext Transfer Protocol Secure, or HTTPS (it appears as a padlock icon in your search bar), which set out rules for sharing data between websites and browsers. These two measures gave people the confidence to start buying things online and it is no coincidence that both eBay and Amazon started in the same year.

The opening up of the internet for financial transactions, the increased use of credit cards, the proliferation of email, the arrival of social media in 2004, and the introduction of Bitcoin in 2009 which allowed anonymous transfers of money, created an ideal environment for cybercriminals. If misanthropic young men could access systems for kicks, they could also access them for money.

The first major online theft was in 1994. Customers of Citibank noticed that money was missing from their accounts. Someone had accessed the bank's computerised cash management system after stealing user passwords and transferred more than $10 million into overseas accounts. The hacker was Vladimir Levin, a Russian student, who was later lured to London where he was arrested and extradited to the US, where he was sentenced to three years in prison. Levin's was a rare form of attack and banks have since put in place much more robust security measures.

A more common form of online crime is credit card fraud. This normally involves hackers accessing credit card databases held by major retailers and selling the information to other criminals who use the details for fraudulent payments or purchases. One of the earliest major online heists was an attack on the US retailer, TJ Maxx, in 2005 by Cuban American Albert Gonzalez, who accessed the credit card data of 45 million users over an 18-month period. He was eventually arrested and more than $1.6 million in cash was seized. He was sentenced to 20 years in prison.

Online credit card fraud can be lucrative, but with spending limits on cards, it takes repeated transactions over a period of time to earn a decent living. Quicker and bigger money can be made using ransomware. This denies access to a system until a ransom, usually in untraceable Bitcoin, is paid. One of the first major attacks was known as *Wannacry* in 2017. *Wannacry* used a worm to target systems and within days, thousands of businesses, and organisations, including the UK's National Health Service (NHS) had been hit, locked out of their own computers unless they paid $300 to regain access. Cyber security experts were able to stop the spread of *Wannacry* after a few days, but enormous disruption had been caused, not least to the NHS which had to spend £70 million on restoring IT systems.

Wannacry was just the highest-profile ransomware attack of 2017. During that same year, the FBI's Internet Crime Complaint Center logged 1,793 ransomware attacks in the US that cost victims over $2.3 million, and many more went unreported. Most of the demands were in the region of $300–500. Low enough for people to pay quickly, but not high enough for the authorities to investigate thoroughly.

Ransomware requires specialist coding skills. These specialists recognised that they could make more money if they developed *ransomware as a service* (RaaS). This is where expert developers produce ransomware which is sold to users for a one-time fee, a subscription service, or a commission based on successful attacks.

The barriers to criminality were lowering. In the old days to rob a bank you needed a gun and bags of courage. Then to be a hacker you needed a computer and coding expertise. Now you just need to be able to download software and identify a target. You had some chance of getting away with a stick-em-up bank robbery, but according to the US think tank Third Way, only 0.3% of all reported cybercrime complaints result in prosecution. That's 3 out of every 1000 crimes. And only a small proportion of crimes are reported. So, a hacker with little skill can commit sophisticated computer crimes, pretty much with impunity

RaaS allowed online robbery on an industrial scale. One of the most prolific RaaS is known as Cryptowall. Between 2011 and 2015 it was used to extort more than $300 million from business and individuals, according to the non-profit US-based Cyber Threat Alliance.

RaaS essentially turns cyberspace into a criminal free-for-all where all devices are constantly under attack from artificial intelligence enabled malware that roams around incessantly exploiting vulnerabilities. In the Victorian age, as we saw in chapter 2, the Brahmah safety lock withstood all attempts to pick it for 67 years. Now cyber security locks can be exploited within months, creating a constant cycle of innovation and exploitation, as cybercriminals chip away at defences.

The business model for ransomware was fast nickels rather than slow dimes. It involved hitting a high volume of users for relatively low sums of money. But the model shifted in May 2021 when hackers went for the big bucks of a single high-value target: the Colonial pipeline, which supplies much of the East Coast of the US with petrol, forcing the closure of 9,500 gas stations.

It was a brilliant strategy, because the cost of the closure of this critical national infrastructure for a day was greater than the $4.4 million cost of the ransom. This was an incentive for a quick resolution and the story made media headlines, adding to the pressure. "I know that's a highly controversial decision," said Colonial CEO Joseph Blount, as he admitted paying the ransom. "I didn't make it lightly. I will admit that I wasn't comfortable seeing money go out the door to people like this."

A month later, in a surprise announcement, the FBI claimed to have recovered $2.3 million of the ransom in a virtual wallet used by Russian hackers. The FBI remained quiet about how they achieved this, but it stands as a rare example of law enforcement success against foreign-based cybercriminals.

The Colonial attack demonstrated a level of sophistication, technical ability and information gathering more normally associated with nation states. Nation states have long had a close interest in cyberspace. It was a vector through which intelligence could be exposed. The first was in 1988 when 16-year-old German Markus Hess hacked into US government computers and sold secrets to the KGB. From then on, cyberspace joined land, sea and air as an active battleground for national rivalry.

State cyber activities remained unseen until the publication in 2013 by US cyber security firm, Mandiant, of a report implicating the Chinese People's Liberation Army (PLA) in what they called, "a multi-year, enterprise-scale computer espionage campaign." Mandiant labelled a PLA Unit 61398, based in Shanghai, as Advanced Persistent Threat Number 1, or APT1.

The term APT means a non-opportunistic group, normally state-run or sponsored, which breaches organisations on a strategic, long-term basis, with clear objectives—normally to steal intellectual property or sensitive information. In other words, they use cyberspace as a vector for espionage, and they are not confined to China.

As well as espionage, states also use cyberspace to attack facilities. The classic example is the 2010 US-Israeli *Stuxnet* attack on an Iranian nuclear facility which caused physical damage to centrifuge systems. As well as physical damage, states use cyberspace to spread disinformation, or fake news, to influence public opinion. There was evidence that Russia used these tactics to influence the outcome of the 2016 US election.

In future conflicts any state that dominates cyberspace is likely to win, so they continuously test each other's capabilities and defences. "Cyber-attacks," says Richard Fenning, former CEO of Control Risks, "are the cheapest form of warfare since the bow and arrow." Furthermore, they don't spill blood, and have fewer consequences for the attacker, especially if they are carried out by a deniable proxy. It is for all these reasons that the British military announced in 2020 that it was trading in tanks for techies and setting up a Joint Cyber Force that will include offensive capabilities.

Hacktivists

Soon after criminal hackers, groups of social activists—hacktivists—appeared on the scene. The most prominent was Anonymous, which, as the name suggests, conceal their identities, while ironically, campaign to stop censorship on the internet. They are a loose international collective that targets a range of institutions including the Church of Scientology, government departments, ISIS, and major corporations. They use what are known as denial-of-service attacks, to lock out users, or post prank messages or videos.

In 2015 another hacktivist group, calling itself The Impact Team, attacked the Canadian dating site Ashley Madison. The group provided an early and high-profile example of what became known as *doxing*: dumping someone's personal information on the internet to embarrass them and harm their reputation.

Membership of a dating site would not normally cause red faces, but Ashley Madison cater for a particular demographic: it encourages married people to have affairs. The names, addresses and telephone numbers of millions of members were splashed by the Impact Team, not for financial gain, but to close down what they saw as an immoral site. Ashley Madison survived, tightened up its security and, strangely, the publicity generated by the attack helped boost its membership. But not before many of the doxed had had very difficult conversations with their partners.

The Impact Team may have affected relations between couples, but another hacktivist group, Wikileaks, affected relations between nations. In the words of its controversial founder Julian Assange, "WikiLeaks is a giant library of the world's most persecuted documents. We give asylum to these documents, we analyse them, we promote them, and we obtain more." In essence, WikiLeaks gained unauthorised access to classified documents and publishes them. Amongst these were quarter of a million diplomatic cables, details of US military activities, and information that damaged Hillary Clinton during the 2016 election campaign. From pranks, to malicious damage, to criminality, hackers could now embarrass governments and affect the outcome of elections. The more the world became dependent on computers and the internet, the more vulnerabilities could be exploited by hackers.

Digilantes

Not all hacking is illegal or malicious. As well as cybercriminals or Black Hat hackers, there is a breed known as White Hats, or ethical hackers, who probe cyber security networks for weakness. These are then disclosed to their owners—in return for what are known as bug bounties. Many big firms including Dyson, Dropbox, and Snapchat, have formal bug bounty programmes that pay thousands of dollars for identifying weaknesses in their security. For some hacking, is a hobby that nets a few hundred dollars a month, for others it's a lucrative career.

The ethical hacking company HackerOne has a platform that matches hackers with companies. In 2019 the platform earned $40 million in bounties and claimed that nine of its hackers netted more than $1 million. Facebook has run a bug bounty programme since 2010 awarding more than 1,500 hackers from 107 countries bounties of up to $80,000 each for identifying security vulnerabilities. In 2021 Microsoft paid out $13.7 million to bug hunters. This was three times as much as in 2018, a sure sign of the increasingly serious nature of cyber security.

Even the US Department of Defense sponsors hackers. Since 2016 it has hosted an annual Hack the Pentagon event to identify vulnerabilities across its cyber defences. Every year hundreds of hackers compete for a bounty pot of $110,000. That seems like a reasonable amount, until you consider that in 2019, according to the FBI, business and individuals in the US lost $3.5 billion to cybercriminals. So it is much more profitable to wear a black hat rather than a white hat.

Regulating Cyberspace

There are surprisingly few laws that regulate the use of cyberspace. It's a bit like the pioneering days of motoring before the highway code, road signs, or driving licences. The key principle, that unauthorised access of systems was a criminal offence, was established in the early days of the IT revolution when the US passed the 1986 Computer Fraud and Abuse Act, and the UK, never far behind, passed the 1990 Computer Misuse Act.

Despite the massive surge in online crime, the numbers convicted are tiny. There are several reasons for this. A lot of cybercrime goes unreported as the sums lost are often relatively small, the victims may be embarrassed to admit that they had been duped, and there is little confidence that funds lost will be recovered, or that cybercriminals will be prosecuted. The last point is understandable: in the UK fewer than 50 people a year are prosecuted under the Computer Misuse Act, while the Crime Survey for England and Wales showed that there were 1.7 million computer misuse offences in 2020.

Also, it's hard to obtain reliable evidence of a crime having been committed. For instance, a digital log showing unauthorised access can be copied and manipulated so it may not withstand cross-examination in court. And a major factor is that many cybercriminals attack victims in other countries. It's hard enough to secure a conviction where victim and criminal are in the same country, but nearly impossible if they are in different jurisdictions. Even if the evidence is strong, there may be no extradition agreement. For example, Russia will not extradite suspects to the US or the UK, so criminals there can operate without jeopardy.

GDP Arrggghhh

As convicting cybercriminals is so difficult, the emphasis is placed on protecting personal data, such as names, addresses, phone numbers, bank details, health records, and even pizza topping preferences. This was the aim of the European Union's 2018 General Data Protection Regulation (GDPR), that had a profound effect on cybersecurity. GDPR is actually a lot more interesting and influential that it sounds. And with maximum fines set at 4% of a company's global turnover it had real teeth.

British Airways (BA) discovered the hard way how impactful GDPR was when hackers accessed its booking system containing the personal details and credit card numbers of thousands of customers. BA reported the incident to the Information Commissioner's Office. There was no evidence that anyone suffered loss, or that any credit cards were used fraudulently. But BA was still fined 1.5% of its global turnover: £184 million (later reduced to £20 million when the pandemic downturn was taken into account).

The Information Commissioner, Elizabeth Denham, said: "People entrusted their personal details to BA, and BA failed to take adequate measures to keep those details secure. Their failure to act was unacceptable and affected hundreds of thousands of people, which may have caused some anxiety and distress as a result."

Fined £184 million for an incident that resulted in no loss, but *may have caused some anxiety and distress*! This was a dramatic change in approach. Like the duty of care cases described in Chapter 12, the protector rather than the perpetrator takes the hit. Imagine a shopkeeper being burgled, reporting it to the police, and then being fined for not adequately securing the stock, while the police do virtually nothing to catch the criminals. That is essentially what GDPR does.

One could argue that modern concerns about data misuse have their origins in Nazi Germany where census data was used to identify Jews, and in post-war East Germany where the Stasi secret service eavesdropped, snooped, and kept meticulous records on everyone to identify subversives. The extent of Stasi activities became apparent after reunification in 1990. The West Germans were appalled at this unauthorised and sinister use of data. They have been suspicious of data collection ever since, becoming data protection zealots, and the godparents of European GDPR.

It is sometimes said, with some truth, that the US aims to monetise data, China aims to control it, Russia aims to weaponise it, and the EU aims to protect it.

The GDPR directive runs to 56,000 words (almost as long as this book). To save you reading it, the key points are about consent and protection. People must give explicit consent for their data to be used. This is why we have all those annoying privacy policy statements that pop up when we are online. The upside is that organisations can only use your data for the purposes that you gave it for. So, if you book a holiday online, the company cannot pass your information to a sun lotion manufacturer for a sales pitch before you go, unless you clicked your agreement.

GDPR does not state what security standards are needed, it just says that organisations must take 'appropriate technical and organisational measures' to ensure that data is protected from unauthorised access. It is similar to the duty of care principle (described in chapter 11) where *reasonable care must be taken against reasonably foreseeable harm*. This means that GDPR can be applied flexibly. A bank and a bookstore can apply different standards of security reflecting the difference in asset value. But for each, the potential to be fined 4% of global turnover is a mighty incentive for taking security very seriously.

Although GDPR is a European law, it has influence well beyond Brussels. If non-European companies want to hold data on European citizens, they need to comply with it, so essentially any company operating globally needs to align with the legislation. GDPR is the highest standard for data protection, compliance is a badge of quality, and it provides confidence to users. In the absence of other legislation, companies such as Facebook have adopted it as their global standard for information assurance.

Securing the Wild Web

Cyber security is bookended by two key principles: unauthorised access of systems and data is illegal, and organisations must protect data. In between there are some industry-specific (such as finance and health) cyber regulations, and there are some general standards (such as ISO27001 or the US National Institute of Standards and Technology), but much of the space in-between remains ungoverned and unregulated.

In the real world, there are comprehensive laws and police crackdown on offenders and protect the vulnerable. In cyberspace there are no police. Criminals operate with virtual impunity and individuals and organisations are obliged to provide their own security. The cyber security environment has parallels with the Wild West in the US, where laws were acknowledged, but order and justice were in short supply. In the Wild West, bounty hunters went after bad guys. In the cyber Wild West, they look for holes in defences. In both cases, the cost of security was and is paid by companies and individuals.

To help offset the costs of cybercrime, cyber insurance became available in the early 2000s. It was a good business to invest in. By 2020 the global market was worth $7.8 billion and, according to US analysts MarketsandMarkets.com, it is forecast to grow to $20.4 billion by 2025. The increase is due to the accelerating pace of sophisticated and costly cyber-attacks, the need to comply with GDPR regulations, the increased remote working during the pandemic, and the recognition that computer networks are the jugular of modern life.

Insurance companies do what they do best: reduce the likelihood of pay-outs. If they are to compensate for cyber losses, they want to ensure that cyber security is in place and kept updated. So cyber insurance pushes up security standards and reduces the risk of a successful attack, whilst covering the costs of any losses.

To manage cyber security, organisations needed posses of their own. Not calloused cowboys on horseback, but soft-handed geeks on padded chairs who understood the difference between fishing and phishing. Their emphasis is not on catching criminals, but on what is known as CIA: confidentiality, integrity and availability. This means that they protect confidential networks, preserve the integrity of data, and ensure that networks and data are available when needed. Their tools are encryption, firewalls, anti-virus software, detection systems, access control, and back-ups. And they spend a lot of time making sure people use long and complex passwords, that are hard to remember, and shouldn't be written down.

While physical security had centuries to mature, cyber security is only two decades old, and rides in tandem with information technology's rapid evolution. Microsoft Windows, for example, has been through 10 major versions and dozens of upgrades in that time. Cyber security was forced to keep pace with the progression.

This evolution reflects Moore's Law which states that the number of transistors on an integrated circuit doubles every two years, which means that computers become more powerful and cheaper, demanding constant software upgrades. And this means that hacking and cyber security technology maintain a rapid loop of constant innovation. In physical security, developments take place over decades; in cyber security, developments take place over months. It's like the difference between travelling on a bicycle or a motorbike.

These days CEOs are becoming laser-focused on cyber security. It's not just about losing data, but also the breach of customer trust, effect on brand reputation, and ultimately, share price. Facebook's dropped from $185 to $152 and usage of the site dropped by 20% in the weeks following the 2018 Cambridge Analytica scandal. Both recovered within 6 months, but it was a clear warning of the impact of user data being accessed without authorisation.

Cyber security programmes are now business critical, and a ransomware attack or a data breach could affect a company's viability. Cyber security has therefore developed an urgency beyond anything experienced within physical security.

As a consequence, there has been a rapid demand for cyber security specialists to work both in house, and in the growing range of specialist companies such as Fortinet, Norton and Darktrace. Traditional security companies such as G4S, Gardaworld, and Control Risks now provide one-stop services that cover both cyber and physical security. Other sectors have also spotted an opportunity to diversify into cyber security including the professional service company PWC, the tool company Stanley, and defence contractor BAE.

Not enough Geeks

The market for cyber security specialists is red hot, but the demand has not been matched by a supply of qualified people. Computer science degrees have been around since the 1980s, but specialist cyber security degrees only appeared in around 2000. By 2020 there were more than 1,500 Bachelor's and Master's level courses in the US and the UK; despite this, there remains a skills shortage. This is reflected in cyber security salaries: in the US the median pay, according to the Bureau of Labour Statistics, is $103,000, much more lucrative than facing down physical threats where guard's median pay is $31,000. And there is no wearing a uniform or standing outside in all weathers.

Until about 2018 any security management job advertised was obviously in physical security. By 2020 a security management job was invariably in cyber security. The market changed as companies scrambled to hire specialists to help them become GDPR compliant. There was no memo sent to inform security professionals of the new reality; cyber security jobs simply swamped the sector and took over the title "security", without even using the prefix *cyber*. On Linkedin, the professional network site, by 2020 you had to assume that all security jobs advertised were in cyber security unless it said otherwise. In the space of two years, physical security had become the junior partner in this universe

The writing had been on the wall for some time. Addressing a conference of international security consultants in 2007, US security consultant Kevin Coleman told the audience with remarkable prescience, "your security career is over as you know it. ... the transition from gates, guns and guards to information, intelligence and integration is technology centric not people centric."

At the senior manager end, Cyber Security Chiefs started to get paid more and were more likely to report to the CEO than their Physical Security counterparts. The talk turned to *convergence*: the integration of physical and cyber security management functions. Both sides protected against loss and both had risk as a common currency. There was also a degree of co-dependence: many physical security functions (access control, intruder alarms and CCTV) relied on IT, and IT systems still needed to be physically secured. A kumbaya moment was imminent—right?

Not quite. The big cultural difference between physical and cyber security is difficult to bridge. Physical is real and defined, cyber is nebulous and as vast as human imagination. Senior physical security managers tend to be ex-police or military Boomers (born before 1965 who grew up with ballpoint pens and polaroid cameras and have a static worldview). Senior cyber security managers tend to be Millennials (born after 1985) or iGeneration (born after 2000); digital natives who grew up with computers and smartphones and who are more comfortable with looking at a screen than looking someone in the eye. Melding the two cultures and generations is a challenge.

Cyber security is a massive and growing global issue. The threats from malicious, criminal, activist, and military users are mushrooming. Cracking down on them, in the vast anonymous world of cyberspace, is extraordinary difficult. It's like an enormous game of *Wackamole*, played in the dark, with the moles getting faster and more numerous every time you hit them. The result is an emphasis on protection, which means the bunkerisation of networks and zero trust policies. Getting into a network becomes increasingly cumbersome, demanding multiple-factor authorisation (password, text, biometrics, and card readers) at every turn. In the real world, a state's first duty is to protect its people. In the cyberworld the state cannot protect. You are on your own. It's a good time to check your passwords, privacy settings, and antivirus software.

15

HOMO SAPIENS TO HOMO SECURITAS

"It's tough to make predictions" said baseball-playing philosopher Yogi Berra, "especially about the future." But that shouldn't stop us from trying. Whether predictions are right or wrong they either impress or amuse, so they are worth a punt. The curious thing is that, even when you can draw upon centuries of history that brought us to this point in time, it is difficult to see what lies even a few years ahead. Considering the future is like launching a stone with a catapult. You stretch back the elastic as far as you can and then let go, only to see your stone fall, uselessly, at your feet. You can look back over 10,000 years, but it's hard to project forward even 10 years with any accuracy.

Even if predictions proved correct, progress is not uniformly smooth. It is like spikes on a sea urchin protruding from an uneven shell. For example, the front end of security has always been provided by low-skilled guards. In most of the world that remains the case, but in advanced societies progress has accelerated to the point where security is provided by high-tech systems, smart CCTV, digital X-Rays, and GPS tracking. You can argue that the future has already arrived, but it's just unequally distributed.

Robocop

The first industrial revolution, starting in the 1780s, brought steam power. The second, from the 1870s, brought electrical power. And the third, from the

1960s, brought automation and computing. In the 2000s we commenced the fourth industrial revolution of cyber-physical systems with digitisation, connectivity, mobile devices, cloud computing, big data, smart sensors, the internet of things, and artificial intelligence. Each revolution brought new technology, but this latest is a catalyst for complete transformation.

Imagine you arrive at work in your car. As you get out a cheery voice greets you by name and says, "Good morning! Remember your ID card and be sure to lock your car." You look around and you see a robot. Its exterior is smooth white plastic, dotted with cameras and sensors. It's about five feet (160 cm) tall and at 400 pounds (180kg) it's no lightweight. It has four holonomic wheels, which means that each is steerable, like the wheels on a shopping trolley. It is shaped, depending on your imagination, like a bullet, or half an elongated egg. Or a phallus.

This is the K5 security robot, first produced in 2015 by Silicon Valley based Knightscope, one of the leading companies in automated surveillance, whose modest aim is to "make the USA the safest country in the world." Designed for use in large private facilities such as factories, data centres, airports, universities, hospitals, and car parks, the K5 is a mobile surveillance platform with microphones, cameras, GPS, sonars, facial recognition, and night vision devices, light detection and ranging sensors, licence plate scanners, and

The future of security: never late, never blinks, never flirts.
(Courtesy of Knightscope)

human temperature monitors. It can operate indoors and outdoors, continually sending data back to a Security Operations Centre (SOC), which is analysed instantly to provide what's known as situational awareness.

The tech website ZDNet suggests that Knightscope is "packaging the bumbling earnestness of mall cops with still imperfect capabilities like autonomous navigation and human machine interaction." But security robots' functionality is continually improving, and the market for them is increasing by 8% a year. It is easy to see why.

Security work is largely routine, repetitive, low-skilled, occasionally dangerous, and it demands long and anti-social hours, so it's ripe for automation. And robots are easier to manage than humans. They don't text whilst working, get stuck in traffic and turn up late, take holidays, catch colds, need a pension plan, or flirt with colleagues. They are cheaper too. You can't buy a K5, you can only rent one. At between $4 and $9 an hour, they undercut minimum wages in the US which varies depending on the state between $8 and $14 an hour. This is a good time to be a robot.

The Drone Ranger

As well as robots with ground surveillance capabilities, drones are increasingly used to provide aerial surveillance for large facilities such as construction sites, mines, ports, logistics bases, and for monitoring major events such as football matches and concerts. Like their ground-based counterparts, drones are fitted with a range of sensors that beam real-time situational awareness to a SOC.

The purists insist on calling them UAVs—Unmanned Aerial Vehicles—but in a world with too many three-letter acronyms, most prefer just to call them drones. The term originated in the 1930s amongst the British scientific community that developed the De Havilland 82B Queen Bee, a radio-controlled target aircraft used for anti-aircraft gunnery training. In homage to the Queen Bee, subsequent models were referred to as drones and the name stuck. The inside joke is that drones were invented so that geeks could get airborne too.

From target practice in the 1930s, the military started to use drones for reconnaissance in the 1960s, and for remote attack in the 1990s. From the mid 2000s the availability of cheap, compact electronics spurred the development of commercial drones for flood and wildfire assessment, and for pipeline and agricultural surveys. By 2013 drones became available for personal use and by 2020, in the US alone, almost 900,000 drone licences had been issued.

The technology progressed so rapidly that by 2021, Ring, the CCTV company purchased by Amazon, developed a small drone to autonomously patrol your home, either on a pre-programmed flight path, or in response to noise or movement. It searches for intruders, and alerts you to windows left open, or the cooker being left on. Retailing at around $250 it opens a new dimension of domestic surveillance capabilities. And puts your dog out of a job.

If there is a bump in the night, you can stay in bed while your drone investigates and beams images to your smartphone. For some it will provide peace of mind. For others it will induce paranoia as they constantly check to see how things are around their home. For everyone, it will normalise surveillance, making people more accepting of the use of drones by the authorities.

Soon after their introduction, recreational drones presented a security headache. In many countries they were used to supply drugs and mobile phones to prison inmates, and this remains a persistent problem.

In 2013, one landed within a few metres of German Chancellor Angela Merkel, as she spoke at a political rally in Dresden, igniting concerns about the potential for their weaponisation. It wasn't long before the threat materialised. ISIS armed off-the-shelf drones with grenades to attack targets in Iraq and Syria in 2016–17. They were more of a tactical nuisance than a strategic threat, but they enabled ISIS to project their influence well beyond their front lines without having an air force of their own.

In Venezuela in August 2018, two drones armed with explosives attacked a military parade attended by President Maduro. He was unharmed, but seven soldiers were injured. In the same year, Greenpeace activists crashed a Superman-shaped drone into a French nuclear power station near Lyon to, in the words of their spokesperson, "highlight the extreme vulnerability of this type of building." And in November 2021 the Iraqi Prime Minister, Mustafa al-Kadhimi, survived a drone assassination attempt in Baghdad's green zone. In 2022 Ukraine's critical infrastructure wasn't so fortunate as the Russians launched hundreds of drones at the power supply network.

Countering the threat is a huge challenge. Drones are small, fast, and difficult to distinguish from birds, making them hard to detect and destroy. Detection technology such as radar, radio frequency, acoustic, or infrared locators is expensive and only partially effective. Even if a drone is detected, taking it out of the sky presents another problem, not least because it may crash onto someone.

Electronic jammers can work, but they often stop legitimate radio signals such as WIFI and mobile communications, so they are seldom an enduring solution.

There are bazooka-type devices that launch nets to trap drones, but they are short-range and easily evaded. Eagles have been trained to attack drones, but keeping them ready to deploy at short notice is rarely sustainable. And physical barriers such as nets may help keep drones away from prisons or nuclear facilities, but they are seldom practical in more conventional settings.

The use of drones is certain to increase dramatically in the future. A 2020 report by the Indian market intelligence firm Coherent Market Insights, claimed that the safety and security drone market was worth $350 million in 2019 and is set to grow at 35% a year for the next ten years. But the initiative will remain with criminals. How long before they use drones to enter homes via open windows to remove valuables, and homeowners start to weaponise their own drones to counter them, sparking home drone combat? It would make a fascinating test of Second Amendment rights in the US. And how long before terrorists succeed in using a swarm of killer drones to attack political targets? Surely, it's only a matter of time.

Smart SOCs

Amongst many magical new technologies that form part of the fourth industrial revolution, lies AI—artificial intelligence. Back in 1981, Steve Jobs described computers as a "bicycle for the mind." AI is like a bicycle for the computer. In 2018, Google CEO, Sundar Pichai, put it more ambitiously, "AI is more profound than fire and electricity." That's a big claim, but when you consider its future potential, he may have a case.

AI is recursive, which means that it instructs itself to analyse data, behaviour, images, and language. It is both a good teacher and a good student. It works 24/7, remembers everything forever, and never stops learning. A human can spend perhaps eight hours a day absorbing information, and then remembers only a portion of it, as you'll recognise when you struggle to recall the French verbs that you learnt at school.

AI uses algorithms, which are essentially a set of rules set in computer code, to understand situations and react logically to them. You have probably used AI several times already today: getting directions on your smartphone, looking at the personalised advertisements delivered to your social media account, asking Alexa to order washing powder, buying theatre tickets from a talking customer service bot, or watching a movie recommended for you on Netflix. AI crept unannounced into our lives. In the future it will be omnipresent, and it will be hard to distinguish AI from human interaction.

A smart SOC provides constant monitoring.
(Courtesy of DSOC)

In chapter 8 we saw how many organisations have developed sophisticated SOCs to control their internal environment, and to monitor the external environments for extreme weather, demonstrations, terrorist attacks, road crashes, flight delays, criminal activity and health hazards, that can impact travel, supply chains, markets, and share prices.

Traditionally, organisations were alerted to events by staff on the ground, or by keeping an eye on the TV news. Increasingly, SOCs use AI to mine data from thousands of news outlets, social media channels, and key websites, to warn of events. If a hurricane is forecast to hit Florida, or if there is a demonstration in downtown Nairobi, or if there is a pandemic surge in Italy, AI can identify all assets, staff, suppliers, and customers likely to be affected, and send them automated alerts specific to each location.

In the SOC, paper is replaced by pixels. Real-time images and data are shown on digital walls that provide situational awareness with a granularity and a speed that the Pentagon would admire. In this age of smartphones people now expect their employer to provide rapid information about security issues affecting them whether they are commuting to work, travelling for work, and, following the rise in remote working during and after the COVID19 pandemic, around their home too.

For employers, meeting these expectations is a sign of corporate virtue and adaptability. Security teams are increasingly involved not only in loss protection, but in organisational resilience, helping to maintain productivity. An agile security team, with the ability to anticipate the future, will increasingly be viewed as an essential source of competitive advantage.

RoboSOC

The cathedrals of our age are smart structures such as data centres, shopping malls, office blocks, and airports. More machine than building, they are assembled with spanners rather than trowels. They are, like spaceships, sealed environments: everything within them—lighting, air quality, temperature, access, parking, occupancy, fire safety, energy management, connectivity, and security—is controlled, not by people pulling levers and flicking switches as they monitor gauges that rise and fall, but by murmuring banks of AI-enabled computers. Any person would be drowned by the tsunami of information that constantly flows from wireless sensors, but AI sails happily on this sea of data.

The rivers of security information generated within; CCTV images, intruder alarms, access control, licence plate scans, ID card data, human temperature sensors, occupancy monitors, robots, and drones would overwhelm a traditional SOC, so that too is AI-enabled with human operators only needed to maintain an overview and resolve ambiguous situations. We are at the dawn of an unnerving new reality where the security of humans in these smart structures is delivered by computer-controlled robots. The ultimate case of the tail wagging the dog.

Imagine that the robot that identified you in the car park detects that you are running a fever. It informs you that you have a temperature of 38°C (100°F) and asks you to leave and seek medical attention. It then sends a message to your smartphone confirming your temperature, informs your line manager that you won't be in today, and updates your HR records. Your access rights to the building are suspended until you upload a medical certificate showing that you are well enough to return to work. Sorted.

Or imagine that you left a bag unattended in a meeting room. It is spotted by smart CCTV which then rewinds and identifies you using facial recognition, and then sends you a message asking you to retrieve it. Or if a door alarm sounds and, instead of sending a security guard to check it, AI automatically reviews the images to see if it was an intruder or you sneaking out for a crafty cigarette. All that happens quicker than you read this sentence.

Smart security systems deliver constant monitoring and immediate decision making. They identify a problem and initiate a solution before a SOC operator has taken a sip of coffee. AI is more observant, quicker, and cheaper than humans. It is not prone to complacency, it doesn't need a shift change or a social media hit, it constantly improves, and never asks for a pay rise.

Inevitably there will be bumps on the road to fully smart systems. Before full automation can occur, an organisation needs to have effective operating procedures that can be programmed into AI. My Facebook example (Chapter 13) showed how the system placed a distressed woman seeking help in the "violent behaviour" category. Impeccable logic led to fallible judgement when what was needed was an understanding of idiosyncratic speech and an empathetic human approach.

But AI will learn from its experience and develop linguistic and cultural dexterity. A new field known as AE—Artificial Empathy—is emerging. A computer's camera and microphone can be used to analyse facial expressions, and tone of voice, to understand how students react to online lessons, and how shoppers interact with automatic checkouts. Think of AI as the brain and AE as the heart. A robot combining the two will have meaningful interaction with humans.

The distinction between security robots, and flesh and blood security guards, is blurring. In the 19th century, some guards were issued with *watchclocks*, devices that mechanically verified that they had done their patrol. In the 1970s these were upgraded with portable electronic data collection devices to place against magnetic tags around the premises. The next step, from about 2010, was for security guards to wear GPS monitors and body cameras that constantly send real-time information to a SOC. These capabilities are now incorporated into smart glasses, essentially a voice-activated smartphone that you wear on your nose. Humans are becoming sensor platforms, and subordinate to computers.

Smart glasses technology has been around since 2012 when Google Glasses were released to the public. They functioned well, but people objected to being constantly videoed by wearers. Some bars and restaurants banned their use, they soon became uncool, and users became known as *glassholes*. In 2021 Facebook teamed up with RayBan to re-launch the technology, banking on the sunglass manufacturers' fashionable brand, and on a shift in public expectations of privacy, they hoped to gain enduring acceptability, rather than a rude nickname.

The continuing advantage of human guards is that they are just that—human. They smile, they greet you by name, they talk to you about the weather, or the ball game. Their presence is reassuring for staff, visitors, and customers. But in the same way that humans are becoming sensor platforms, robots are becoming more human. They can interact with people, provide directions, and issue safety instructions.

They also offer a solution to dealing with violence. Rather than have a human guard handle an aggressor, you can use a robot with facial recognition to identify them and, using its empathetic interactive speech capability, negotiate with them. If all else fails, the robot could use a Taser to incapacitate them until the police arrive. It's not as far-fetched as it sounds. In 2016, Dallas police deployed a robot, normally used for bomb disposal work, to kill a gunman who had shot five police officers.

The concept of robots killing humans was the theme of the 1984 Arnold Schwarzenegger movie The Terminator, about a cyborg (a robot with a human exterior) sent on an assassination mission by an AI system. With remarkable prescience, the film was set in 2029, by which time real life AI and robotics will have many of the capabilities of the fictional Terminator, and robots will have replaced thousands of guards.

Data Points

If, rather than observing everything with CCTV, everything has an electronic tag, its location would be known all the time. Static visual observation is replaced by dynamic digital monitoring, and you can account for everything that happens, everywhere, all the time.

Take the example of cars. Speed limits were first policed by officers hiding behind lampposts with handheld radars. The next step was static automated enforcement cameras that flashed as you raced by. These freed up police time and caught greater numbers of speeding drivers, but it was still possible to exceed speed limits on roads without cameras. However, if you want drivers to obey the law all the time, fit cars with an onboard telematic device that constantly monitors them and sends the data to a central computer that automatically issues fines.

Many car insurance companies offer policies that are priced according to driver behaviour that is monitored using a telematics device. Drive like a girl (really.... they are less prone to risky driving—and one of the leading companies in this field is drivelikeagirl.com) and you can expect a 10–20% reduction in the cost of your premium. Drive like Lewis Hamilton chasing a championship, and you can expect a price hike. The emphasis is on rewarding people for driving safely, which provides an incentive to use the technology.

Studies show that when drivers know that their driving is being monitored, their accident rate drops by 20%. Every year, worldwide, more than 1.3 million people are killed on the roads. A global rollout of these devices has the potential to save hundreds of thousands of lives a year. This is the point where privacy and safety collide. Whilst we may instinctively loathe the idea of monitoring, intellectually we can appreciate the wider benefits.

I told you that, to tell you this: personal security is taking a similar route to road safety. We are on the cusp of swapping CCTV observation for continuous dataveillance. Everyone who uses a smartphone is essentially electronically tagged; everyone is a data point that is continuously monitored (described in Chapter 6).

Your location is triangulated using the network of mobile phone repeater masts and your phone's GPS facility. This information is available to your data provider and can be accessed by the police. Your location can also be gauged through your social media accounts, along with your search history, contacts, and photographs. We know that we are revealing much about ourselves online but, at every turn, the benefits outweigh the costs. We'll return to this theme in a few pages when we consider what's next.

Cybergeddon

The key tools of the future physical security world—robots, drones, telematics, and AI-enabled SOCs—all rely on cyberspace for their existence. Yet cyberspace, as we saw in chapter 14, is also a new arena for criminals and hackers. In England and Wales in 2020/21 police figures show that there were 267,931 victims of domestic burglary compared with 1.7 million victims of cybercrime, such as hacking and malware. Fewer than 10% of burglaries resulted in convictions. Less than 0.003% of cybercrime results in a conviction. It's a low-risk, high-reward occupation, for both geeky criminals, sophisticated criminal groups, and state actors, all of whom see limitless potential in the wild west of cyberspace.

In chapter 4 we saw how a crime wave surged from the 1960s because of the rise in portable consumer goods, increased drug use, loosening social cohesion, social inequality, and lack of effective policing. We now see a surge in cybercrime due to a rise in the use of online systems, underdeveloped cybersecurity, more intellectual property being stored digitally, the rise of digital financial assets, the borderless nature of the cyber environment, the ability to remain anonymous online, and the very low prospect of being caught. The wave of cybercrime will get much bigger before it is brought under control.

Signing an Executive Order on cyber security in May 2021, President Joe Biden said, "The United States faces persistent and increasingly sophisticated malicious cyber campaigns that threaten the public sector, the private sector, and ultimately the American people's security, and privacy." Biden followed up in August 2021 calling for a national cyber security effort including developing an encryption culture, more cyber insurance linked to security standards, the training of an army of cyber security specialists, and enlisting the help of the tech giants Apple, Google, and Microsoft, to provide cash and expertise.

In 1953 President Dwight Eisenhower set up *Project Solarium* to consider how to respond to a nuclear-armed Soviet Union. In 2019, in recognition of a threat of similar magnitude, The Cyberspace Solarium Commission was established to "develop a consensus on a strategic approach to defending the United States in cyberspace against cyber-attacks." This is recognition of the US's cyber vulnerabilities, and a belief that WWIII will be fought in cyberspace with highly unpredictable outcomes. For the US, the ability to dominate cyberspace in the future will be as important as dominating the land, sea, and air in the past.

Put Your Heart Into It

Let's leave aside the chilling prospect of WWIII and return to the concept of people being data points. As a security manager, I'm interested in three things: who you are, where you are, and how you are. I was involved in deploying diplomats and aid workers to war zones in the early 2000s. When travelling, they would report their location, via crackling radio, to a SOC where they were tracked using pins on a paper map.

In 2004 the company Track24 revolutionised the scene with a device that tracked vehicles' locations using satellites and "pinged" coordinates at regular intervals to a SOC that could observe them on an electronic map. It also incorporated a panic button that would send an alarm, and turn on a recording device, so that SOC operators could track people and listen to what was happening around them. Because they were priced by satellite usage, they were only set to "ping" every few minutes to limit costs. Compared with occasional radio reporting they seemed like magic at the time and worth the extra money. Track24 also produced a personal "pocket buddy" version, although many diplomats and aid workers were reluctant to use something that also revealed where they were, and who they were with, even when off duty.

The arrival of smartphones in 2007 opened the market for Track24-type capabilities using apps that worked through the mobile phone network rather than expensive satellites. SOCs in conflict zones started to switch to locator apps, and diplomats and aid workers started to lose their inhibitions about sharing their data.

In 2017 the social media platform Snapchat introduced Snap Map, which shares, with consent, your location. Suddenly its 300 million teenage users had a level of situational awareness that most military commanders could only dream of. Or have nightmares about. In November 2017 the fitness app Strava published maps of its user data saying, "Our global heatmap is the largest, richest, and most beautiful dataset of its kind. It is a direct visualisation of Strava's global network of athletes." It included data from serving military personnel who used it to track their personal fitness regimes. Some were serving in Syria, Iraq, and Afghanistan and their running routes, around secret bases, were illuminated online for all to see.

Older generations, for whom a telephone was in a box at the end of the street that ate a pocketful of coins every few minutes, are instinctively wary of tracking devices. Generation Z, those born after 2000, who grew up online, are much more relaxed about them. In recent years dozens of tracking apps and watches have hit the market, promising greater safety and security for everyone. What not to like?

The simplest trackers just provide the location of an individual. More advanced versions are aimed at families and have a range of features. For example, *FindMyKids* records all movement, sends alerts if a geofence is breached, includes a panic button, allows parents to see what apps their kids are using, what calls they make, and even listen in to their conversations. In a few short years the technology migrated from the battlefield to the playground.

Kids can now be monitored more closely than electronically tagged criminals on parole. This inhibits a kid's ability to navigate life independently. But the technology is cheap, convenient, and it may be helpful in a critical situation. Your child is the most precious thing in your world: if you have a chance to provide a little additional security, why not? The alarming aspect is that future generations will grow up believing that this is normal. And it will be.

The boundaries of *normal* are being stretched. Since the beginning of time, babies have cried to signal their distress. Now there's an app for that. Once confined to hospital intensive care units, the technology is available for parents to track their "child's heart rate, oxygen level, and sleep trends from the first night you bring them home to their first day of school." Called the Owlet smart sock, it contains sensors that measure vital signs, and sends the data to your smartphone, "so," as Owlet's website explains, "while they rest easy, you can too." But do you rest easy, or do you constantly check your phone to see if your baby is still alive? Physiology now becomes subordinate to technology and the consequences will be profound. Let me explain.

In chapter one we discussed how our bodies are essentially the same as those of our ancestors who lived on the African savannah. We react automatically and involuntarily to an impending car crash, in the same way as they did to an approaching lion. Our fear mechanism starts with the amygdala, we become hyperalert, our pupils dilate, our breathing accelerates and our heart rate immediately spikes. Many activities can quicken the heart, but only shock causes it to accelerate from a resting beat to 200 beats per minute in an instant.

Heart rate monitors are also a useful tool to help with the growing issue of mental health. Our hearts reflect not only our immediate fear, but also our gnawing anxiety. If you have a resting pulse of 65 beats a minute, but after a few months of working in a new, high-pressure job, it has increased to 85 beats per minute, it is likely that you are stressed. Heart rates can also reveal how kids are doing at school, if they are being bullied, or if they are taking drugs. A child might insist "I'm ok Dad," but their heart rate data can reveal if something's up. As a parent, wouldn't you want to know?

Once infants grow out of the Owlet their parents, hooked on the ability to constantly check on them, may insist that they graduate to a heart tracker watch, or a family finder app. The same parents are likely to be amongst the growing number of people who, following a trend started by elite athletes, track their own health and fitness using heart rate monitors produced by companies such as Fitbit, Garmin, and Apple.

The technology is spreading beyond medics, parents, and fitness enthusiasts, into professional life. In chapter 12 we saw how the concept of duty of care was broadening. Employers are obliged to take *reasonable measures against reasonably foreseeable harm* or risk civil litigation. Heart rate monitors are becoming part of the package of *reasonable measures*.

The Cambridge-based company Equivital produces a device called the *Lifemonitor* to measure vital signs and analyse the data of people working in hazardous and stressful occupations, such as firefighting, deep-sea diving, soldiering, and emergency medicine. It comprises an undervest with sensors that measure heart and breathing rates, core temperature, location, and body position (so it can tell if you have fallen over). It comes with a software package that plots people and their data on a map and it analyses their performance, like a Premier League footballer. If anyone is stressed, exhausted, lost, entering a particularly risky area, or injured, they can be identified immediately.

Every breath you take, every move you make, we'll be watching you.
(Courtesy of Equivital)

As you can imagine, the *Lifemonitor* provides an extraordinary ability to manage people's safety. But once it becomes standard equipment for front-line emergency services who face physical harm, the case is likely to build for second-line workers who are exposed to stressful situations, such as those who drive fire trucks, or handle 999 calls. And from there, the case builds to monitor the vital signs of pilots, intensive care staff, and others in safety-sensitive functions. Armed with this data, employers can optimise individual performance, review working practices, and arrange medical intervention. Human performance can be fine-tuned, like a machine.

It might sound like the future, but it has already arrived. In September 2022 bus drivers in Beijing started a trial of similar sensors that identify if they are unfit for work through illness or because of a poor emotional health. It seems like an egregious breach of privacy, but in 2016 a tram crashed in Croydon, London and killed 6 people. It transpired that the driver fell asleep at the controls, probably due to him regularly sleeping less than 6 hours before waking up at 3am to go to work. With a *Lifemonitor* his high-risk lifestyle might have been identified and the crash prevented.

You will recall from chapter 6, that Amazon purchased Ring, the home surveillance company described by the Guardian in 2021 as, "the largest corporate-owned, civilian-installed, surveillance network that the US has ever seen." Alphabet, which owns Google, also owns Nest (which has a similar CCTV product range to Ring) along with Fitbit. And Facebook is getting into surveillance imagery via its collaboration with RayBan.

These companies have intimate knowledge of you via your online data, and a growing knowledge of your off-line activity through CCTV and heart rate monitor feeds. Google, which also owns YouTube, can, in theory, track your heart rate as you watch a movie or view an advertisement, to understand your emotional engagement and to provide you with even more precisely targeted content. The other big incentive for tech companies to get into heart monitoring, is that they open the door to the health sector which is set to expand massively as we get evermore focussed on wellness and eternal life.

Heart rate monitoring as a security tool, leapfrogs facial recognition. If you monitor everyone's heart-rate and location, facial recognition becomes redundant, and it opens the door to a new realm of security. If everyone was obliged to share their data with a State-run and AI-enabled SOC, what would happen? Inevitably, rather like drivers with telematics, behaviour would improve, crime would fall, and everyone would be more secure. If you were in trouble, say you fell over while walking in the woods, the system could dispatch help to you. If you had a temperature, you could be ordered to self-isolate. If you were attacked, your assailant would be immediately identified, the digital evidence would be hard to refute, and justice would be swiftly served. Once everyone was connected to the system, resources could be shifted from protective security to instant response.

Is this vision of the future an Orwellian dystopia or a secure utopia: Big Brother, or Big Mother? The Big Brothers see it as sinister totalitarianism, handing control to regimes that would inevitably use it to quash dissent and preserve their own power. Big Mothers would say that if you are doing nothing wrong, you have nothing to fear. Surrendering privacy is worth it, if it means eliminating security threats and controlling ever expanding and more complex societies.

While the prospect of the State exerting control through location and heart rate monitoring seems unthinkable, the technology to do so exists already, and at every opportunity people show a remarkable willingness to gift their data in return for more security.

Staying Alive

The boundaries between security, health, and well-being are blurring. Ultimately, they are all about staying alive, which is not only a catchy tune by the Bee Gees, but also the prime motive of every animate being. We were designed to instinctively evaluate the threats to our lives on the African savannah (lions, snakes, cliffs and lightning), but we are not equipped to evaluate new dangers such as flying, radiation, coronavirus, crossing the road, eating doughnuts, or terrorism. Technology provides us with the ability to navigate modern life and stay alive, even longer.

Life expectancy is a function of three things: our DNA, our environment, and our lifestyle. We are fast developing the tools to manage all of these. For under $100 we can get a DNA test that will indicate our risk of getting cancer, psoriasis, or Parkinson's. Data on environmental risk (such as air quality, transport hazards, and homicide rates) are increasingly available. And our life-style risks (diet, exercise) can now be measured with some accuracy. Data helps us to manage our lives and to set our own risk thermostat. Ultimately it will enable us to set a dial to live slow and die old, or live fast and die young.

A statistical concept known as a micromort can be used to measure your personal risk. The micromort was introduced by Stanford Professor, Ronald A Howard, to express your likelihood of dying during any particular activity. A micromort is a one millionth chance of accidental death, which equates to about 30 minutes of your overall life expectancy.

For example, micromorts can help inform your travel choices; you use one micromort for every six miles travelled on a motorcycle, or every 250 miles travelled by car (so motorcycling is over 40 times riskier than driving). Or your choice of sport: you always knew that climbing Everest was hazardous, but if you knew it used 38,000 micromorts per summit would you opt for a parachute jump instead which uses only ten micromorts?

Smoking 20 cigarettes a day uses 10 micromorts, being sedentary (i.e., watching TV) for two hours a day uses 1 micromorts. In other words, every day that you smoke reduces your life expectancy by about 5 hours, and watching two episodes of Game of Thrones reduces your life expectancy by 30 minutes.

The good news is that Cambridge Professor David Spiegelhalter introduced the concept of microlives, a measure by which you can extend your life by 30 minutes: eating well (plus 4 microlives for a balanced diet) or exercising (plus 2 microlives for 30 minutes of exercise). And, in a triumph of oestrogen over testosterone, women gain 4 microlives a day more than men as they are better at managing risk.

Online you can find various sites such as *Deathclock* that use data to predict when you will die. They are both morbid and fun, but they underline a serious point: your lifestyle choices affect your life expectancy, and they can be measured with increasing granularity.

And you can take out telematic travel insurance. If you opt for a safer destination, avoid travelling by public transport, stay away from sketchy parts of town, and make it back to your hotel before dark, you get a lower premium than if you stay out all night in dive bars in a high crime neighbourhood. Security can become personal, quantified, and monetarised, as long as behaviour is monitored.

The concept is both thrilling and appalling. It might make us more secure, but as we end up using algorithms to make decisions, we erode our natural instincts. Are we inevitably heading for a future of bovine contentment, wired up to AI devices, part human, part machine, our emotions neutered and manipulated? Or does it mean that we jettison our savannah instincts that are disorientated by the modern world, and replace them by tech systems adapted to the new reality? The path that we are on seems inescapable, and given its destination, there is surprisingly little public debate or pushback.

Faust and Franklin

The term *security* refers to both precautions to guard against harm, and to an end state where you are secure. So, security is both the journey and the destination. In the same way that a state of true peace means not just the absence of war, but a sense that war is inconceivable, a state of absolute security means not just absence of threats but a sense that harm is inconceivable. That will be the point at which our appetite for security will be finally satiated.

In the future people are likely to use the term *security* in the context of its Latin roots and seek a life that that is *se*—without, *cure*—concerns. In the future, the key performance indictor of any security programme will be measured in heart beats per minute.

The technology of the fourth industrial age, and our instincts for survival, force us into a Faustian pact where we trade privacy and liberty for security. It would outrage Benjamin Franklin who famously said: "Those who would give up essential liberty, to purchase a little temporary safety, deserve neither liberty nor safety." But what technology offers is not temporary, it's a profound and enduring shift in our ability to manage our concerns.

Franklin suggests there is a binary choice between liberty and security. The reality is much more complex. Without security nothing thrives, and no one is content. Security provides a platform for liberty. We can reduce crime, and we can get help the moment that we get into trouble, but in return we must lead transparent lives, and surrender our most personal data. The choice we make is more between privacy and security, than liberty and security.

The irony is that, in the developed world, we have never lived in more secure times, yet we continue to strive for yet more. As I hope this book makes clear, there is no simple explanation for this. Technology clearly plays an important role, so too does the growing concept of legal liability. Money is also a big driver. Security companies stoke our concerns about loss and soothe them by providing protective services. Insurance companies sell us policies while limiting their own liabilities. The media use fear-based stories to sell papers, clicks and airtime. And politicians know that, to stay in power, their first responsibility is to protect people. They would never be elected by telling us that we are safe enough. There was a time when a state had to defend you from other countries, but now it must guarantee your safety within your own.

Perhaps the biggest factor is our own nature. We always want more security because we set every improvement as a baseline for even more. Security has become a new religion that asserts itself with increasing urgency, even as our lives are better and longer than ever before.

We are on an unstoppable journey from Homo sapiens to Homo securitas.

SELECT BIBLIOGRAPHY

Books

Adams, John, Risk, London, UCL Press, 1995.

Ansary, Tamim, The Invention of Yesterday: a 50,000 Year History of Human Culture, Conflict, and Connection, New York, Public affairs, 2019.

Badr, Christopher D, Baker, Joseph O, Day, L. Edward, and Gordon, Ann, Fear Itself: The causes and consequences of Fear in America, New York, NYU Press, 2020.

Beck, Ulrich, Risk Society: Towards a New Modernity, Sage, 1992.

Bernstein Peter L, Against The Gods: The Remarkable Story of Risk, John Wiley and Sons, 1996.

Bettmann, Otto L, The Good Old Days—They Were Terrible! New York, Random House, 1974.

Blastland, Michael, Spiegelhalter, David, The Norm Chronicles: Stories And Numbers About Danger, Profile Books, 2013.

Bourke, Joanna, Fear: A Cultural History, London, Virago, 2005.

Brundage, Fitzhugh W. Civilizing Torture: An American Tradition, Harvard University Press, 2018.

Button, Mark, Doing Security Critical Reflections and An Agenda For Change. Palgrave MacMillan, 2008.

Churchill, David, Janiewski, Dolores, and Leloup, Pieter (eds), Private Security and the Modern State: Historical and Comparative Perspectives, Abingdon: Routledge, 2020.

Dartnel, Lewis, Origins: How the earth Made Us,The Botley Head, 2018.

Davis, Mike, City of Quartz: Excavating the Future in Los Angeles, New York, Verso Books, 1990.

Diamond, Jared, Guns Germs and Steel: A Short History of Everybody for The Last 13,000 Years, Random House, 1998.

Johnson, Les, Shearing Clifford, Governing Security: Explorations in Policing and Justice, London, Routledge, 2013.

Hamilton, John T, Security: Politics Humanity and the Philology of Care, Princeton University Press, 2013.

Hess Karen M. Introduction to Private Security 5th edition, Cengage Learning, 2009.

Ip, Greg, Foolproof: Why Safety Can be Dangerous and Danger Makes Us Safe, London, Headline, 2014.

Fennelly Lawrence J, Tyska, Louis A, Beaudry Mark H Security in 2020, ASIS International, 2010.

Furedi, Frank, Culture Of Fear Risk Taking And The Morality Of Low Expectation, Cassell, 1997.

Gardner, Dan, Risk: The Science And Politics Of Fear, Virgin Books, 2009.

Gill, Martin, Ed., The Handbook of Security, second edition, Basingstoke, Palgrave MacMillan, 2014.

Glassner, Barry The Culture of Fear Why Americans Are Afraid of the Wrong Things, Basic Books, 1999.

Goode, Erich, and Ben-Yehuda, Nachman, Moral Panics: The Social Construction of Deviance, Wiley-Blackwell, 1994.

Harari, Yuval Noah, Sapiens: A Brief History of Humankind, Vintage, 2015.

Harari, Yuval Noah, Homo Deus: A Brief History of Tomorrow, Vintage, 2017.

Hogg, Garry, Safe Bind Safe Find The Story Of Locks Bolts And Bars, New York, Criterion Books, 1968.

Horton, Alex, Gregg, Aaron, Use of military contractors shrouds true costs of war, Washington Post, 30 June 2020.

Isenberg, David, Shadow Force: Private Security Contractors in Iraq, Praeger Security International, 2008.

Jeffery, Clarence Ray, Crime Prevention Through Environmental Design, Sage Publications, 1978.

Koerner, Brendan, The Skies Belong to Us: Love and Terror in the Golden Age of Hijacking, Broadway Books, 2014.

Krahmann, Ele, States, Citizens and The Privatisation of Security, Cambridge University Press, 2010.

Laquer, Walter, The New Terrorism: Fanaticism and the Arms of Mass Destruction, Oxford University Press, 2000.

Lipson, Milton, On Guard: The Business of Private Security, New York, Quadrangle, 1975.

Loader, Ian, Walker, Neil, Civilising Security, Cambridge University Press, 2007.

Martin, Paul, The Rules of Security: Staying Safe in A Risky World, Oxford University Press, 2019.

McFate, Sean, The Modern Mercenary: Private Armies and What They Mean for the World Order, Oxford University Press, 2014.

Morris, Ian, War: What is it Good For? The Role of Conflict in Civilisation, from Primates to Robots, Profile, 2014.

Moss, Kate, Security and Liberty Restriction by Stealth, Palgrave MacMillan, 2009.

Mueller, John, Overblown: How Politicians and the Terrorism Industry Inflate National Security Threats, and Why We Believe Them, Simon and Schuster, 2009.

Mueller John and Stewart Mark G Terror Security and Money Balancing The Risks Benefits And The Costs Of Homeland Security, Oxford University Press, 2011.

Nemeth, Charles, P, Private Security and the Law, 4th edition, Routledge, 2012.

Peltier, Heidi, The Growth of the "Camo Economy" and the Commercialization of the Post-9/11 Wars, "20 Years of War," a Costs of War initiative based at Boston University's Pardee Center for the Study of the Longer-Range, 30 June 2020.

Pinker, Steven, The Better Angles of Our Nature, New York, Penguin Books 2011.

Pike, Luke Owen, History of Crime in England, illustrating the changes of the laws in the progress of civilisation, London, Smith, Elder and Co. 1873.

Ropeik, David, How Risky Is It, Really? Why Our Fears Don't Always Match the Facts, McGraw Hill, 2010.

Scahill, Jeremy, Blackwater: The Ride Of The World's Most Powerful Mercenary Army, Nation Books, 2007.

Schneier, Bruce, Carry On: Sound Advice From Schneier On Security, Indianapolis, Wiley, 2014.

Scott James C, Against The Grain A Deep History Of The Earliest States, Yale University Press, 2017.

Sheehan, Michael A, Crush the Cell: How to Defeat Terrorism Without Terrorizing Ourselves, Crown, 2009.

Singer. P. W., Wired for War: The Robotics Revolution and Conflict in the 21st Century, Penguin 2011.

Singer P. W., Corporate Warriors: The Rise of the Privatized Military Industry, Cornell University Press, 2007.

South, Nigel, Policing for Profit: The Private Security Sector, Sage 1988.

Talib, Nassim Nicholas, The Black Swan: The Impact of The Highly Improbable, Penguin, 2013.

Vince, Gaia, Transcendence: How Humans Evolved Through Fire, Language, Beauty and Time, Penguin Random House, 2019.

Wrangham Richard The Goodness Paradox How Evolution Made Us Both More And Less Violent, Profile Books, 2019.

Zedner, Lucia, Security, Routledge, London and New York, 2009.

Articles and Reports

Duren Banks, Hendrix, Joshua, Hickman, Matthew, Kyckelhahn, Tracey, National Sources of Law Enforcement Employment Data, U.S. Department of Justice, Office of Justice Programs, Bureau of Justice Statistics, April 2016.

Beckett, Charlie, Fanning The Flames: Reporting Terror in A Networked World, Tow Center for Digital Journalism, 22 September 2016.

Blakely, Edward J, Fortress America, Northern Architecture, 17 December 2020.

Button, Mark, The 'New' Private Security Industry', the Private Policing of Cyberspace and the Regulatory Questions, Journal of Contemporary Criminal Justice, Vol 36, 4 December 2019.

Churchill, David, Security And Visions Of The Criminal: Technology, Professional Criminality And Social Change In Victorian And Edwardian Britain, BRIT. J. CRIMINOL. (2016) 56, 857–876.

Churchill, David, The Spectacle of Security: Lock-Picking Competitions and the Security Industry in mid-Victorian Britain, October 2015, History Workshop Journal 80(1).

CONTEST: The UK's Strategy for Countering Terrorism, June 2018.

Cruikshank, George, Stop Thief: Hints to Housekeepers to prevent Housebreaking, Pamphlet, 1851.

Epstein, Susan B. Diplomatic and Embassy Security Funding Before and After the Benghazi Attacks, Congressional Research Service, September 10, 2014.

Feldstein, Steven, The Global Expansion of AI Surveillance, The Carnegie Endowment For International Peace, 17 September 2019.

Frost, Peter, Harpending, Henry C, Western Europe, State Formation, and Genetic Pacification, Evolutionary Psychology, 13(1), 2015.

Graham, Stephen, CCTV: The Stealthy Emergence of a Fifth Utility? Planning Theory and Practice 3(2), January 2001.

Harari, Yuval Noah, The Theatre of Terror, The Guardian, 31 Jan 2015.

Holman, Alison E, Garfin, Dana Rose, and Cohen Silver, Roxane, The Media's role in broadcasting acute stress following the Boston Marathon bombings, PNAS January 7, 2014 111(1) 93–98.

Claus, Dr Lisbeth, Duty of Care of Employers for Protecting International Assignees, their Dependents, and International Business Travelers, International SOS, White Paper 2009.

Congressional Research Service, Securing U.S. Diplomatic Facilities and Personnel Abroad: Legislative and Executive Branch Initiative, December 23, 2014 (R43195).

Jacobs, Sam, Bubble-Wrapped Americans: How the U.S. Became Obsessed with Physical and Emotional Safety, 2019, Ammo.com.

Jenkins, Simon, Letting Terror Win, The Spectator, 9 April 2016.

Kissenger, Henry, How the Enlightenment Ends: Philosophically, intellectually—in every way—human society is unprepared for the rise of artificial intelligence, The Atlantic, June 2018.

Kitteringham, Glen, Environmental Crime Control, The Professional Protection Officer, IFPO, 2010.

Kopel, David B, The Posse Comitatus And The Office Of Sheriff: Armed Citizens Summoned To The Aid Of Law Enforcement, Journal of Criminal Law and Criminology, Vol 116, Fall 2016.

Krahman, Elke, Security: Collective Good or Commodity? European Journal of International Relations, Volume 14, Number 3, 2008, pp.379–404.

Lea, Robert, BA Fined a Record £183m After Hackers Stole Customer Details, The Times, 9 July 2019.

Loeffler Jane C. Embassy Design: Security v Openess, Foreign Service Journal. September 2005.

Lofstedt, Ragnar E, Reclaiming Health And Safety For All: An Independent Review Of Health And Safety Legislation Presented to Parliament by the Secretary of State for Work and Pensions by Command of Her Majesty, November 2011.

Moss, E, Burglary insurance and the culture of fear in Britain, c. 1889–1939, Historical Journal 54(4), 2011.

Moules, Danny, A history of hacking and hackers, ComputerWeekly.com, 25 October 2017.

Neocleous, Mark, Security, Commodity, Fetishism, Critique, December 2007.

Parfomak, Paul W. Congressional Research Service, Guarding America: Security Guards and U.S. Critical Infrastructure Protection, November 12, 2004.

Perlroth, Nicole, Cyber Warfare And State Sponsored Hackers-The Next Global Crisis, The Times, 25 February 2021.

Provost, Claire, The Industry of Inequality: Why the World Is Obsessed with Private Security, May 12, 2017, The Guardian.

Reuters, Movie Theater Chain Cinemark in Court Over 2012 Aurora Massacre, May 9, 2016.

Richie, Hannah, Hasell, Joe, Appel, Cameron, Roser, Max, Terrorism, OurWorldinData.org July 2013. revised in November 2019.

Ritchey, Diane, Proud To Be Security: How Roles Changed After 9/11, Security Magazine, 1 September 2011.

Routley, Nick, The Crime Rate Perception Gap, VisualCapitalist.com, 5 February 2019.

Rosenbaum, Ron, Secrets of the Little Blue Box, Slate, 1971.

Saunders, The Hon Sir John, Chair, Manchester Arena Inquiry Volume 1: Security for the Arena, Report of the Public Inquiry into the Attack on Manchester Arena on 22nd May 2017, June 2021.

Schmidt, Eric, U.S. Takes Steps to Add Security at Embassies, New York Times, 20 May 2013.

Serani, Deborah, If It Bleeds, It Leads: Understanding Fear-Based Media, Psychologytoday.com, 7 June 2011.

Shearing, Clifford and Stenning, Philip, Modern Private Security: Its Growth and Implications, in Crime and Justice, Vol 3 (1981), The University of Chicago Press, 1981.

Simpson, John, Hackers Avoid Jail Despite Rise in Attacks, The Time s, 27 December 2016.

Smith, David, L, Under Lock and Key: Securing Privacy and Property In Victorian Fiction And Culture, PhD Dissertation, Nashville, Tennessee, August, 2007.

South, Nigel, Private Security and Social Control: The Private Security Sector In The United Kingdom Its Commercial Functions And Public Accountability, PhD Thesis, Middlesex Polytechnic, July 1985.

Stimson Study Group, Protecting America While Promoting Efficiencies and Accountability, May 2018.

Strauss, Neil, Why We're Living in the Age of Fear, This is the safest time in human history. So why are we all so afraid? Rolling Stone, 6 October 2016.

Strom, Kevin, Berzofsky, Marcus, Shook-Sa, Bonnie, Barrick, Kellie, Crystal Daye, Horstmann, Nicole, and Kinsey, Susan, The Private Security Industry: A Review of the Definitions, Available Data Sources, and Paths Moving Forward, U.S. Department of Justice, Document 232781, December 2010.

Sukel, Kayt, Beyond Emotion: Understanding the Amygdala's Role in Memory, Dana Foundation, 13 March 2018.

Supreme Court, New York County, New York, IN RE: World Trade Center Bombing Litigation, Decided: January 20, 2004.

Tiersky, Alex, Epstein, Susan B, Securing U.S. Diplomatic Facilities and Personnel Abroad: Background and Policy Issues, Congressional Research Service, July 30, 2014.

United States Department of State, Bureau of Diplomatic Security, DSS Then And Now: The First Century Of The Diplomatic Security Service, Updated.

United States Department of State, The Inman Report: Report of the Secretary of State's Advisory Panel on Overseas Security, 1985.

United States Government Accountability Office, Report to Congressional Addressees, GAO-17-681SP, DIPLOMATIC SECURITY: Key Oversight Issues, September 2017.

US Government Printing Office, The Security Failures of Benghazi, Hearing before the Committee on Oversight and Government Reform, 10 October 2012, Serial No. 112–193.

Ward, Jaap, D, The Private Security Industry in International Perspective, European Journal on Criminal Policy and Research 7(2):143–17, January 1999.

Wiatrowski, William J, On Guard Against Workplace Hazards: Security Guards Face A Variety Of Workplace Hazards That Can Lead To Injury, Illness, Or Death, US Bureau of Labour Statistics, Monthly Labor Review, February 2012.

White, Dr Jessica, A Complex Matter: Examining Reporting on Terrorism in the UK, RUSI, 4 March 2021.

Ylimaunu, Timo, et al, Street mirrors, surveillance, and urban communities in early modern Finland, Journal of Material Culture, Volume 19, Issue 2, 2014.

WEB RESOURCES

https://www.9-11commission.gov

https://www.asisonline.org

https://bigbrotherwatch.org.uk

https://www.bls.gov

https://www.comparitech.com

https://www.cpni.gov.uk

https://cyberthreatalliance.org

http://www.deathclock.com

https://www.economist.com

https://www.gapminder.org

https://historyofinformation.com

https://icoca.ch

http://www.historyoflocks.com

https://www.historicallocks.com

http://privacyinternational.org

https://www.theisrm.org

https://www.lightbluetouchpaper.org

https://www.gov.uk/government/organisations/national-counter-terrorism-security-office

http://psm.du.edu

https://securityconference.org/en/

https://ourworldindata.org

https://www.professionalsecurity.co.uk

https://www.securitymagazine.com

https://www.securitydegreehub.com

https://security-institute.org

https://www.schneier.com

https://www.tsa.gov

http://understandinguncertainty.org

https://www.visualcapitalist.com

https://en.wikipedia.org

https://www.wired.com

https://www.zdnet.com

Movies

Airport, George Seaton, 1970.

Dark Night Rises, Christopher Nolan, 2012.

Jaws, Steven Spielberg, 1975.

Minority Report, Steven Spielberg, 2002.

Bourne Supremacy, Paul Greengrass, 2004.

Route Irish, Ken Loach, 2011.

The Terminator, James Cameron, 1984.

War Games, John Badham, 1983.

The Wild Geese, Andrew V MacLeglan, 1978.

BIOGRAPHY

Mike Croll was a uniformed guard at a supermarket before going on to hold some of the most senior appointments in international security management including with the Foreign and Commonwealth Office, the European Union, the United Nations, and with Facebook.

Along the way he has also been a diplomat, an army bomb disposal officer, a humanitarian deminer, and a university lecturer.

ACRONYMS

ADT	American District Telegraph
AE	Artificial Empathy
AI	Artificial Intelligence
AQ	Al Qaeda
ASIS	American Society for Industrial Security
CCTV	Closed Circuit Television
CIA	Central Intelligence Agency
CPTED	Crime Prevention Through Environmental Design
CRAM	Counter Rocket, Artillery, and Mortar
CT	Counter Terrorism
FBI	Federal Bureau of Investigation
HTTPS	Hypertext Transfer Protocol Secure
ICoCA	International Code of Conduct Association
IED	Improvised Explosive Device (homemade bomb)
GPS	Global Positioning System
IRA	Irish Republican Army
ISIS	Islamic State of Iraq and Syria
JTAC	Joint Terrorist Analysis Centre
MANPAD	Man Portable Air Defence System

MI5	Military Intelligence Section 5, the domestic Security Service
Mi6	Military Intelligence Section 6, the internal Secret Intelligence Service
RSO	Regional Security Officer
SAM	Surface to Air Missile
SIA	Security Industry Authority
SOC	Security Operations Centre
SSL	Secure Sockets Layer
UAV	Unmanned Arial Vehicle (Drone)
VCR	Video cassette Recorder

Security is an acronym rich sector. I've aimed to spare you the full alphabetti soup, but these were unavoidable.

ACKNOWLEDGEMENTS

I was encouraged to write this by the late David Clark, Chairman of ASIS UK, whose boundless good nature, and enthusiasm for the security industry, was always inspiring.

Throughout my career I have met many dedicated professionals who worked in demanding, difficult, and sometimes dangerous situations. They are too numerous to list here, but I'd like to acknowledge their camaraderie and their role in shaping my ideas for this book.

I'm grateful to Professor John Adam for his iconoclastic perspectives over many years and to Professor Martin Gill and Dr David Churchill for their generosity. Thanks also to Andrew Lowie for believing in this project, and to Rob Campbell for his invaluable editing.

I'm indebted Mark Valley, Phyllida Middlemiss, Anne Blanchflower, Barry Duplatis, Paul Jefferscn, Charmaine Parsons, and to Kevin, Emma, and Ben Croll, who were kind enough to review various chapters, and especially to the heroic Howard Harvey who wrangled them all.

Most of all I'd like to thank my wife, Silvia, for her immense contribution to this book, and for always making me feel secure.

INDEX

2
2nd amendment, 33, 108

7
7/7, 21, 53, 91, 98, 100, 103, 105–8, 114, 125–7, 129, 133, 135, 137, 172, 180, 193, 196, 221, 226, 240

9
9/11, x, 46, 65, 91, 96, 98, 103–8, 111–26, 145, 146, 172, 179, 180

A
Aberfan, 64
Abraham Lincoln, 38
Access control, vii, x, 47, 49, 50, 78, 88, 96, 100, 104, 116, 123, 134, 135, 160, 163, 197, 204, 205, 213
Admiral Bobby Ray Inman, 159
ADT, 52, 59, 247
Aegis, 148, 149
Afghanistan, 16, 103, 112, 115, 126, 131, 145, 147–50, 152, 153, 156, 162, 172, 228
Air India, 98, 111, 141, 161
Airport, vii, ix, 48, 49, 94, 96, 100–8, 111, 127, 133, 135, 136, 141, 166, 178, 180, 218, 223, 224
Al Qaeda, 100, 104, 112, 114, 145, 156, 160, 161, 177
Alan Pinkerton, 38
Alarm, vii, x, 13, 14, 26, 46–54, 59, 70, 82, 84, 98, 165, 194, 202, 216, 223, 227, 228
Albert Gonzalez, 207
Alfred C Hobbs, 22–4
Alison Holman, 66
Alistair Morrison, 143, 149
Allied Universal, x, 59
All-seeing eye, 18, 38, 75–90
Amazon, 83, 84, 189, 192, 207, 220, 230
Ambassador, 144, 156–9, 166, 171, 196
Amygdala, 3, 4, 62, 229
Andrew Marr, 64
Andrew Nightingale, 143
Angela Merkel, 220
Anonymous, 210
Armoured vehicles, 34, 59, 149, 166–9
ARPANET, 204
Arthur Conan Doyle, 28
Ashley Madison, 210
ASIS, 42, 154
Assa Abloy, 24
Aurora Colorado, 179
Aviation and Transport Security Act 2001, 105–6

B
Baby Faced Nelson, 52
Baghdad, 147–53, 162, 166, 169, 171, 196, 220
Bank Protection Act 1968, 52

Baron Haussmann, 78
BBC, 63–4
Bee Gees, 232
Beirut, 98, 156, 159–61, 171
Ben Wallace, 131
Benghazi, 171, 172, 241
Benjamin Franklin, 233
Big Brother Watch, 86
Bill Bratton, 46, 47
Bill Gates, 189
Bitcoin, 207
Black Tom Island, 111
Blackwater, 149, 151, 152, 154
Bob Denard, 142, 153
Bodycam, 82
Bonnie and Clyde, 52
Borough Market, 129, 130
Bouncers, 57, 58
Bow Street Runners, 19, 20, 35
Brahmah, 22–26, 208
Brandon Welsh and David Farringdon, 85
Brendan Koerner, 94
Brinks, 34, 59
British Security Industry Association, 81
Broken windows theory, 46, 78
Bug bounties, 211
Burglary, 26, 28, 29, 44, 52, 226
Business continuity, x, 118–123

C

C. Ray Jeffery, 78
Canal Hotel, 162
Capn' Crunch, 205
Captain Edward Walter, 16
Cash in transit, 34, 47, 49, 59
CCTV, vii, x, 13, 17, 47, 49, 50, 55, 76, 77, 80–88, 118, 122–27, 134, 136, 163, 166, 198, 216, 217, 220, 223, 225, 226, 231
Cesare Beccaria, 18, 19
Charles Dickens, 18, 19, 26
Charles Hamilton Houston, 187
Charles Peace, 28
Cheetham cricket ground, 182

China, 82, 87, 88, 209, 213
Chris Stevens, 171
Chubb, 22, 24, 26, 54, 186
CIA, 171, 200
Cohortes Urbanae, 13, 14
Collective security, 9, 10, 18, 20, 32, 78, 90
Colonel Michael A Sheehan, 116
Colonial pipeline, 208
Compstat, 46, 47
Computer Fraud and Abuse Act 1986, 205, 211
Computer Misuse Act 1990, 211
CONTEST, 127
Control Risks, 143, 145, 148, 209, 215
Convention for the Elimination of Mercenarism in Africa, 153
Convention for the Suppression of Unlawful Seizure of Aircraft 1970, 95
Cornhill, 24–26
Corporal punishment, 8, 37
Corporate Manslaughter and Corporate Homicide Act, 182
Corps of Commissionaires, 16, 17, 25
Corps Security, 17
Counter-terrorist spending, 115
CPTED, 78, 79
C-RAM, 189
Crime wave, ix, 28, 43–47, 52, 59, 78, 226, 227
Crisis management, x, 118–23, 185
Cuba, 77, 93, 95, 96, 106, 190, 204, 207
Curfew, 24, 187
Cyber security, x, 203–218, 226, 227
Cybercrime, 203–208, 211–214, 226, 227
Cyberspace Solarium Commission, 227

D

Daily Telegraph, 25
Daniel Mullinger, 185
Dar es Salaam, 160
Dataveillance, vii, 76, 87, 89, 226
David Cameron, 128
David Lee Smith, 206
David Omand, 88

David Pekoske, 108
David Spiegelhalter, 232
David Sterling, 143
David Walker, 143, 145
Dawson's Field, 95, 106
DB Johnson, 93
Deathclock, 232
Department of Homeland Security, 105, 114
Dick Cheney, 114
Donald MacPherson, 175
Donald Trump, 46, 152
Dr Deborah Serani, 66
Dr Who, 204
Drones, x, 83, 219–223, 226, 248
Drugs, ix, 26, 41, 44–5, 48, 106, 129, 158, 220, 226, 229
Duty of care, 173–187, 212, 213, 216
Dwight Eisenhower, xi, 137, 227
Dystopia, 163, 190, 231

E

Edward Snowden, 113
Egypt, 7, 8, 20, 21, 53, 104, 107
Elizabeth Denham, 212
Embassy, x, 100, 141, 144, 147–9, 154–66, 170–72, 184, 185, 200, 238–40
Emma Carr, 86
Emotion, vii, 1, 2, 4, 27, 28, 41, 55, 62–6, 74, 85, 113–15, 121, 131, 134, 164, 230–33, 239
Equivital, 229, 230
Erik Prince, 152
European Convention on Human Rights, 86
Executive Order 13228, 114
Executive Outcomes, 142

F

Facebook, 65, 86, 89, 121, 189–202, 211, 213, 215, 224, 231
Facial recognition, vii, 84–88, 218, 223, 225, 231
Factory Act 1802, 67
Fasces, 37

Faustian pact, 233
FBI, 52, 59, 86, 118, 203, 206, 207, 209, 211
Fear, v, viii, ix, 1–5, 9, 10, 13, 15–21, 23–29, 49, 51, 52, 61, 62, 65, 66, 70, 73, 75, 78, 84, 88, 112, 114–18, 126, 129, 130, 133, 140, 157, 168, 169, 175, 182, 185, 194, 202, 229, 231, 234–40
Figen Murray, 134, 136
FindMyKids, 228
Firewalls, 206, 214
Flak jackets, x, 155
Ford, 39, 40
Ford Pinto, 176
Fourth industrial revolution, 218, 221
France, 8, 26, 78, 100, 133
Frank Sinatra, 91

G

G4S, 59, 149, 150, 215
Gardaworld, 149, 150, 215
Gated communities, 50, 51
GCHQ, 89
GDPR, 212–16
Geeks, 24, 190, 195, 203, 205, 207, 209, 211–15, 219, 226
Geneva Conventions, 153
George Cruikshank, 26
George Orwell, 77, 88, 231
George W. Bush, 105, 112, 114, 146
God, 8, 9, 65, 75, 76
Google, 23, 63, 189, 201, 221, 224, 227, 231
GPS, vii, 76, 88, 121, 217, 218, 224, 226
Great Exhibition, 22
Green zone, 162–65, 169–71, 196
Guard Dog Act 1975, 57

H

HackerOne, 211
Hackers, x, 24, 89, 203, 205, 207–15, 226, 239, 240
Hacktivist, 210
HAL, 202

Hammurabi, 7
Harry Bennett, 40
Health and Safety Executive, 68
Heart, 3, 4, 13, 21, 40, 55, 62, 129, 224, 227–33
Henry Mayhew, 18
Henry Wells, 34
Hijack, ix, 91–99, 101–9, 141, 149, 179
Homestead Steelworks, 38, 59
Homo sapiens, xii, 1, 2, 4, 217, 234
Homo Securitas, xii, 217, 234
Homosexuality, 158
HTTPS, 206
Hue and cry, 9, 32
Huron, 32

I

I Love You, 206
Iain MacLeod, 71
Ian Crooke, 143
ICAO, 96, 100
ICoCA, 154
If it bleeds it leads, 65
Industrialisation, 18, 67, 68, 78
Inman Commission, 159, 160, 162
Insurance, ix, 24, 28, 29, 49, 54, 88, 119, 120, 144, 145, 175, 180–87, 214, 225, 227, 233, 234
Intelligence, xi, 1, 47, 76, 89, 99, 100, 103, 112, 113, 119, 122, 125, 127, 128, 133, 137, 145, 147, 148, 166, 168, 186, 195, 198, 201, 208, 209, 216, 218, 221
International SOS, 185, 186
Intruder detection systems, vii, 17, 52–4
IRA, 81, 85, 120
Iraq, 16, 103, 112, 115, 126, 131, 142, 143, 145, 147–53, 156, 159, 162, 166, 168–72, 220, 228
Istanbul, 47, 126, 130, 161, 162

J

Jack Maple, 45, 47
Jaws, 66
Jean-Jacques Rousseau, 5
Jeff Bezos, 189

Jeremiah Chubb, 22
Jeremy Bentham, 77
Jerusalem, 13
Jimmy Savile, 71
Joe Biden, 227
John Holmes, 179
John J Duncan, 107
John Mueller and Mark Stewart, 133
Jon T. Willie, 76
Joseph Blount, 238
Joseph McCarthy, 41, 158
JTAC, 127, 128, 137

K

K5 security robot, 218, 219
Kabul, 149, 151, 162, 170, 171
Karl Benz, 70
Karl Largerfeld, 69
Keenie Meanie Services, 143
Ken Loach, 153
Kevin Coleman, 216
Kidnap and ransom, 120, 144, 169
King Edward I, 10, 11, 14
Kroll, 149

L

L'Aquila, Italy, 183
Las Vegas, 180
Lawyers, x, 46, 94, 144, 162, 173–87, 194
Lewis Hamilton, 225
Liberty, vii, 7, 86, 90, 106, 109, 111, 112, 189, 233
Lion, viii, ix, x, 1–5, 9, 229, 232
Lock, vii, ix, 3, 6, 7, 11, 19–29, 47, 52, 53, 59, 84, 208
Lockerbie, 116, 117, 118, 145
Locksmiths, ix, 22
London, ix, 5, 14, 16, 18–27, 35, 45, 58, 59, 70, 80–2, 98–100, 126, 127, 130, 131, 133, 135, 139, 141, 143, 144, 158, 161, 194, 195, 199, 205, 207, 230
Lord Palmerston, 27
Louis Freeh, 86
Lt Colonel Robert Brown, 142

M

Mall Cop, 117
Malware, 206, 208, 226
Manchester Arena, 129, 133–136
Margaret Thatcher, 135
Margee Kerr, 66
Marijuana, 48
Mark Twain, 78
Mark Zuckerberg, 189, 191
Maslow, viii, 50
Matthew Broderick, 205
May Donoghue, 181
Mayflower Compact, 32
Media, xi, 65, 74, 81, 89, 90, 111, 112–15, 121, 126, 130, 131, 200, 202, 207, 208, 221, 222, 224, 226, 228, 234
Melissa, 206
Mercenary, 140–42, 153, 154
Mesopotamia, 7
Metal detectors, vii, 96, 102, 134, 136, 179
MI5, 127
Michael Howard, 80
Michael Shermer, 65
Microlives, 232
Micromort, 232
Microsoft, 24, 189, 211, 214, 227
Mike Davis, 49, 57
Mike Hoare, 141
Militias, 11, 32, 33, 38–40
Milton Lipson, 57
Minority Report, 90
Montreux Document, 153
Monty Python, 93
Moral panic, ix, 25–7, 41, 116
Moscow, 157–8
Mr Winterbottom, 173, 175–6
Murder Act, 8

N

Nairobi, 156, 160, 171, 222
Nanny State, 71, 73
Ned Ward, 15
New York, 34, 36, 45–7, 52, 83, 96, 99, 104, 111, 114, 118, 125, 175, 178, 198, 205
Nick Lovrien, 200
Niccla Mendelsohn, 195
Niscur Square, 152, 153
Norbert Weiner, 204

O

Occupational Safety and Health Act in 1970, 68
Onel de Guzman, 206
Osama bin Laden, 65, 112–14, 117

P

Panopticon, 77, 89
Paris, 19, 78, 79, 133
Patriot Act, 113
Paul Bremmer, 152
Paul Maynard, 131
Paul Moxness, 63
PFLP, 95
Philip K Dick, 90
Photoguard, 80
PLA Unit 61398, 209
PMC, 139, 140, 143–54, 166, 171
Police, vii, ix, 10, 13, 19, 20, 24, 25, 27, 31, 36, 37, 40–9, 52, 56, 57, 59, 78, 80–5, 90, 93, 96, 108, 118, 127, 131, 133, 135, 137, 166, 169, 177, 194, 195, 199, 201, 206, 212, 214, 216, 225, 226
Port Authority of New York, 178
Posse, 9, 11, 32, 35, 214
President Maduro, 220
Pretorian Guard, 14
Prison, 16, 18–24, 27, 33, 45, 75, 77, 91, 141, 152, 163, 182, 200, 203, 207, 220, 221
Privacy, vii, 75, 77, 83, 86, 88, 90, 113, 196, 213, 216, 224, 226, 230–33
Private Security Industry Act 2001, 58
Professor David Nutt, 129

R

Ransomware, 207, 208, 215
Red Flag Act 1865, 70
Richard Bethell, 149
Richard Burton, 141

Richard Fenning, 209
Richard Grimshaw, 176
Richard Nixon, 106
Richard Reid, 100
Richard Wrangham, 5
Risk management, x, 119
Road Safety, 71
Robert Morris, 205
Robot, 218, 219, 223–26
Roger Moore, 141
Roger Short, 161
Romans, 7, 9, 37, 140
Rome, 13, 14, 95, 96
Ronald A Howard, 232
Royal Society for the Prevention of Accidents, 70
RSO, 166
Rudy Giuliani, 46, 114

S

Safety, viii, 4, 28, 53, 62, 63, 66–74, 78, 86, 92, 93, 95, 102, 107–09, 114, 115, 121, 137, 172, 176, 177, 180, 182–87, 208, 221, 223, 225, 226, 228, 230, 233, 234, 236
Samuel Johnson, 2, 144
Sarin gas, 202
SAS, 143–45
Secure Embassy Construction and Counterterrorism Act 1999, 161
Securicor, 59
See it, Say it, Sorted, 131
See something, say something, 118, 131
Sergio Vieira de Mello, 161
Shakespeare, 15
Sheriff, 10, 11, 32, 35–7, 166
Sherlock Holmes, 28, 35
Shopping mall, ix, 43, 49, 51, 55, 134
Silicon Valley, 189–95, 200, 201, 218
Simon Jenkins, 128
Sir John Saunders, 136
Sir Peter Imbert, 135
Sir Robert Peel, 20
Sky Marshall, 124, 125
SOC, 122, 123, 198, 199, 219, 221–28, 231
Social credit, 87
Soldiers, vii, 16, 17, 31, 40, 139, 140, 144–47, 150, 153, 169, 184, 220
South Africa, 66, 161, 162
Special Forces Club, 139, 144, 154
Spy mirrors, 77
SSL, 206
Staff Cop, 89
Star Trek, 52
Stasi, 213
State Troopers, 37
Statute of Winchester, 10
Stay safe, 63
Stephen Morrill, 119
Stephen Paddock, 180
Steve Jobs, 189, 205, 221
Steve Wozniek, 205
Steven Pinker, 130
Steven Spielberg, 90
Strava, 228
Stuxnet, 238
Sun Tzu, 131
Sundar Pichai, 221
Support Anti-Terrorism by Fostering Effective Technologies Act, 180
Surveillance, xii, 48, 52, 55, 73, 75–90, 112, 113, 123, 127, 218–20, 231
Sweden, 59, 71
Swiss Guard, 140

T

Take care, 63
Taser, 225
Terrorists, x, 4, 62, 65, 85, 91, 94–6, 100–07, 111–21, 125–37, 149, 158–62, 171, 175, 177–81, 221, 222
The Great Lock Picking Controversy, 22
The Terminator, 225
The Wild Geese, 141
Thomas Hobbes, 2
Tim Berners-Lee, 204
Tim Spicer, 142, 148, 149
Tom Jenkins, 133

Tony Blair, 130, 133
Track24, 227, 228
Trees, 1, 61, 120, 122, 185, 190
TSA, 105, 106, 108

U
Ulrich Beck, 81
United States Guard, 40
US Marshall Service, 35
Usain Bolt, 1
USSR, 41, 157

V
V2, x, 80, 83
Value of a statistical life, 115
Vietnam, 44, 52, 102, 142
Vigiles, 13, 14
Vladimir Levin, 207

W
Wackenhut, 59
Wagner Group, 141
Walter Laqueur, 130
Wannacry, 207
War Games, 205
Watch and Ward, 11, 14
Watchclocks, 224

Watchman, xii, 11, 13–17, 24, 31, 43, 53, 76, 119
Wikileaks, 210
Wilc West, 33, 36, 37, 214, 226
Wilhelm Balck, xi
William Fargo, 34
William Fitzhugh Brundage, 32
William Gibson, 204
William J. Burns, 39
World Trade Centre, 100, 104, 107, 111, 177
World Wide Web, 204
WWI, 40, 72, 125
WWII, x, 6, 41, 80, 92, 106, 125, 141, 143, 147
WWIII, 227

X
X-Ray, vii, 96, 99, 100, 102, 136, 178, 217

Y
Yale, 21, 24
Yellow vests, 69, 70, 72
Yogi Berra, 217
Yuval Noah Harari, 130

Ingram Content Group UK Ltd.
Milton Keynes UK
UKHW021944080523
421401UK00008B/691